HOW BUILDINGS LEARN

HOW BUILDINGS LEARN

What Happens After They're Built

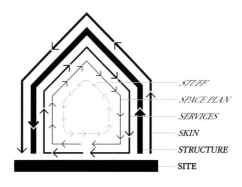

STUFF
SPACE PLAN
SERVICES
SKIN
STRUCTURE
SITE

Stewart Brand

PHOENIX ILLUSTRATED

First published in Great Britain by Viking in 1994
This revised paperback edition published in 1997 by Phoenix Illustrated, a division of
Orion Books Ltd, Orion House, 5 Upper St Martin's Lane, London WC2H 9EA

A CIP catalogue record for this book is available
from the British Library.

ISBN: 0 75380 0500

Printed and bound in Italy

Grateful acknowledgement is made for permission to reprint "The Bed By the Window" from
Selected Poems by Robinson Jeffers. Copyright 1932 by Robinson Jeffers. Copyright renewed
1960 by Robinson Jeffers. Reprinted by permission of Random House, Inc.

This book has been the subject of a legal dispute and certain
sections have been changed in the British edition.

Designed by Stewart Brand and Donald Ryan

Contents

ACKNOWLEDGMENTS

"PORCHES FILL IN BY STAGES, not all at once, you know." The architect was responding to a talk I gave at a builders' conference. "The family puts screens on the porch one summer because of the bugs. Then they see they could glass it in and make it part of the house. But it's cold, so they add a duct from the furnace and some insulation, and now they realize they're going to have to beef up the foundation and the roof. It happens that way because they can always *visualize* the next stage based on what's already there."

I didn't get the architect's name. He stands in my mind for all the ghost collaborators on this book that I've failed to credit properly. Perhaps culture is driven most by just such flea-market ideas in a vast network of uncredited influence. Generic thanks is due to that network.

As for names, I'll begin with the architects, in honor of the abuse their profession catches in this book—much of it quoting them, of course. It will be obvious from the text how much I owe to Frank Duffy*, currently president of the Royal Institute of British Architects, and to Chris Alexander*, author of *A Pattern Language* and professor at the University of California Berkeley (UC Berkeley, hereafter). Less conspicuous is the contribution of Sim Van der Ryn*, UC Berkeley professor and former State Architect of California. He started all this by encouraging me to lead a seminar called "How Buildings Learn" at the UC Architecture Department in 1988. The students in that course, including many working professionals, were early co-conspirators.

Others starring in the text are John Abrams*—design/builder on Martha's Vineyard (an island off Massachusetts)—and Richard Fernau & Laura Hartman* (the ampersand signifying they are a firm, in Berkeley). Behind the scenes but also influential are a crowd of architects and designers—C. Thomas Mitchell* (author of *Redefining Designing*), Tedd Benson* (author of *The Timber-Frame Home*), Gordon Ashby (museum designer), Robin Middleton (Architecture Association, London), Jean de Monchaux (then head of MIT's Architecture Department), Herb McLaughlin (of KMD, San Francisco), John Worthington (of DEGW, London), Coby Everdell, (Bechtel, San Francisco), Bill McDonough (Mr.

Sustainable Architecture), Rodger Messer (CRSS, Houston), William de Cinces (Universal Pictures), Pete Retondo, Allen Roberts, David Sellars, Edward Shoucair, Richard Tobias, and Susan Shoemaker.

Planners had a hand in this, as usual—notably "neotraditional" town planners Peter Calthorpe* (San Francisco) and Andres Duany (Florida), MIT's campus planner Robert Simha, and the author of *Built for Change*, Anne Vernez Moudon. Hands-on builders included contractor Matisse Enzer*, remodeler Jamie Wolf*, Whole Earth's design-maven J. Baldwin*, carpenter and shipwright Peter Bailey, and information-technology consultant David Coggeshall. Even more in the swim of my subject were facilities managers Chuck Charlton (at Stanford University) and Anita Lewis-Antes (at Princeton). My informants on real estate, where the real power is, included Dick York (Sausalito, California), Jonathan Rose (New York), Geoff Woodling (then of Jones Lang Wootton, London), and Nicholas Wakely (of Vail Williams, England).

Building preservationists, time-intoxicated, gave me early encouragement and direction, especially Penny Jones*, Director of Preservation Programs at America's National Trust for Historic Preservation. Other helpful professionals were Ward Jandl (National Park Service) and Russell Kuene (International Council on Monuments and Sites). At every historic building were keepers of the flame generous with their time and knowledge—the Duchess of Devonshire* at Chatsworth (Derbyshire), Lloyd Fairbanks at the Fairbanks House (Massachusetts), Director Neil Horstman and librarian Barbara McMillan at Mount Vernon, Director Christopher Scott and researchers Ann Miller and Larry Dermody at Montpelier, Connie Weismuller at Tor House (California), and Liz Robinson at an old shipyard called Marinship.

My greatest research joy was in libraries, swarming through photo collections, reveling in glorious or obscure tomes. Home base was the UC Berkeley Environmental Design Library, ably run by Elizabeth Byrne. The value of its immense collection was multiplied by being computer-searchable from anywhere. Another trove was the Library of Congress—America's greatest governmental treasure—where Craig D'Ooge (public relations) and Ford Peatross (Prints Collection architectural historian) were invaluable. Betty Monkman guided me through the rich photo documentation at The White House. Arthur Olivas in Santa Fe did the same with the Museum of New Mexico's wonderful collection, well

---

* Names with asterisks performed the generous, onerous, crucial service of reading and critiquing all or part of the next-to-final draft of *How Buildings Learn*.

represented in this book. In Britain, Stephen Croad and Gordon Smith helped me with the huge photo archive of the Royal Commission on the Historical Monuments of England, London; Jill Lever showed me choice items in the Drawings Library of the Royal Institute of British Architects; and Douglas Matthews was gracious at the London Library.

Boston turned out to be a major resource area, thanks to Rodney Armstrong and Sally Pierce of the Boston Athenaeum, Sally Beddow and later Kara Schneiderman at the MIT Museum, and Lorna Condon at the Society for the Preservation of New England Antiquities. I'm grateful also for special assistance from Stan Ritchey at the Historic New Orleans Collection, Al Regensberg at New Mexico's State Records Center and Archives, Robert MacKimmie of the California Historical Society (San Francisco), and Paul Roberts at William Stout Books in San Francisco—a museum-quality architectural bookshop.

Architectural historians, particularly those specializing in vernacular buildings, were essential. Dell Upton* at UC Berkeley went through my photos and helped interpret their mysteries. Christopher Wilson* in Albuquerque dissected the tangled past of Santa Fe and the Southwest. Orlando Ridout V in Annapolis demonstrated how to "read" old buildings. William Seale had the stories behind the stories of Mount Vernon, Montpelier, Monticello, and The White House, and Conover Hunt illumined the Montpelier saga. Other instructors were Jules Lubbock, Paul Groth, J. B. Jackson, Fikret Yegul, Henry Millon, Tony Wrenn, and Robert Irving.

I am indebted to many other writers, notably Clem Labine (founder of *Old House Journal*), Joel Garreau* (*Edge City*), Mira Bar-Hillel (London journalist), Robert Campbell (Boston architectural critic), and Ellen Perry Berkeley. Publishers especially helpful in finding original photos included Lloyd Kahn (Shelter Publications), Cole Gagne (*Old House Journal*), Wayne Bonnett (Windgate), and James Robertson (Yolla Bolly). Among the most supportive photographers were Robert S. Brantley (cover), Art Rogers, Christopher Simon Sykes, Peter Vanderwarker, and Robert Nugent.

The pleasure of research travel was greatly increased, and its cost greatly reduced, by the people who let me stay in their homes—Mary Clemmey (the book's British agent) and later Hannah Eno in London, Danny and Patty Hillis in Boston, and Robert Horvitz, John Petersen, and my late

sister Clare Sampson in Washington, DC. When I needed isolation for writing, Ron and Seejoy Berkowitz had the perfect log cabin.

For the full six years of research, writing, and book production I was immersed in two institutions that generously gave every support—Global Business Network (Peter Schwartz*, president), which employs me, and Whole Earth (Point Foundation) in Sausalito, where I am vaguely emeritus. The team that made the book happen was drawn from both organizations. Donald Ryan* of Whole Earth skillfully managed the whole physical production of the book, did all the paste-up (page and illustration preparation), created most of the drawings and diagrams, helped with design, and coordinated with the publisher and printer. James Donnelly* of Whole Earth inked every page of the manuscript bright red with line-editing corrections, for which I am whimperingly grateful. Christina Gerber (and earlier, Danica Remy) from Global Business Network tracked down photos from books and handled permissions labors. All my photos were printed by Tony Iadavaia, of Photography Unlimited, who has a special touch with Ilford XP2 film.

The original publishing deal was made with Daniel Frank at Viking in New York and Ravi Mirchandani at Penguin in London, through literary agent (and longtime friend) John Brockman*. David Stanford* at Viking was the project's editor, handling everything at the New York end adroitly. Roni Axelrod managed the book manufacture, and freelancer Eli Liss did the index.

I asked three friends to serve as muses for the work—that is, loan me portraits to post near the computer where I wrote and did page layout. I wrote, in a sense, to them—figuring that anything that tickled them would tickle anybody. They were musician/artist Brian Eno*, computer scientist Danny Hillis*, and architect William Rawn*. Of course my main muse up close was my wife Patty Phelan*, who knew when to stay clear of the fray and when to wade in and save the day.

I wish authors would put their addresses in their books. Publishers are ill-equipped to forward mail, and authors need contact from readers to make corrections for later printings and to make things happen in the real world. So, enjoy the book; let me know where it needs repair.

—Stewart Brand, Global Business Network, PO Box 8395, Emeryville, California 94662; (510) 547-6822 FAX: 547-8510

Elevation on St Charles Street

Lot Nº 1.   Lot Nº 2.

4 May 1857. New Orleands Notarial Archives, Plan Book 43, Folio 46. The image appears in *New Orleans Architecture, Vol 2, The American Sector* (Gretna, LA: Pelican, 1984), p. 82. This photo is by Robert S. Brantley, 1993—reproduced with the permission of the Office of the Custodian of Notarial Records for the Parish of Orleans, Stephen P. Bruno, Custodian.

COVER STORY...

1857 - On St. Charles Street in the "American sector" of New Orleans, two identical Greek Revival brick townhouses were built about 1850 on adjoining properties. Their addresses would eventually stabilize as number 822 (left) and number 826 St. Charles (right). They both happened to be up for sale in 1857 when this auction watercolor drawing was made. It showed facades of exposed brick, in the American east coast fashion. The modest ornament had a nice detail: the "denticulated cornice" (row of tiny rectangles like teeth) above the doorway was echoed at the top of each building.

November, 1993. Robert S. Brantley, New Orleans.

TYPICAL CHANGES:

Both buildings grew.

They diverged.

Their skins changed markedly.

TYPICAL FEATURES:

Both had rapid turnover of tenants.

Brick construction helped them last.

Window openings stayed the same.

1993 - Both buildings endured vigorously into the 1990s, gaining character and individuality with every decade. Sometime in the 1860s, shortly after they were built, both acquired cast-iron balconies (called "galleries" in New Orleans) on the front and side. Number 822 (left) kept its, but 826 (right) moved on to other additions in three directions—an attic-story upward, an annex to the side, and expansion of the ground floor out to the sidewalk. Meanwhile 822 added a full story on top. After a few initial decades of stable (separate) ownership and occupancy, each building began to experience rapid turnover of owners and rental tenants. By 1936, both facades had been coated with stucco, scored to look like stone masonry. A 1970s photo of 826 showed a barbershop in the ground floor. Since that photo, the building acquired shutters (presumably influenced by the old auction drawing), lights on top, a metal railing for the annex deck, and French doors on the ground floor. In both buildings only one thing seems to be perpetually renewed no matter what: those original denticulated cornices.

CHAPTER 1

Flow

YEAR AFTER YEAR, the cultural elite of San Francisco is treated to the sight of its pre-eminent ladies, resplendently gowned, lined up in public waiting to pee. The occasion is intermission at the annual gala opening of the opera. The ground-floor ladies' room at the Opera House is too small (the men's isn't). This has been the case since the place was built in 1932. As the women are lined up right next to the lobby bar, their plight has become a traditional topic of discussion. The complaints and jokes never change. Neither does the ladies' room.

Between the world and our idea of the world is a fascinating kink. Architecture, we imagine, is permanent. And so our buildings thwart us. Because they discount time, they misuse time.

Almost no buildings adapt well. They're *designed* not to adapt; also budgeted and financed not to, constructed not to, administered not to, maintained not to, regulated and taxed not to, even remodeled not to. But all buildings (except monuments) adapt anyway, however poorly, because the usages in and around them are changing constantly.

The problem is world-scale—the building industry is the second-largest in the world (after agriculture). Buildings contain our lives and all civilization. The problem is also intensely personal. If you look up from this book, what you almost certainly see is the inside of a building. Glance out a window and the main thing you notice is the outside of other buildings. They look so static.

Buildings loom over us and persist beyond us. They have the perfect memory of materiality. When we deal with buildings we deal with decisions taken long ago for remote reasons. We argue with anonymous predecessors and lose. The best we can hope

for is compromise with the *fait accompli* of the building. The whole idea of architecture is permanence. University donors invest in "bricks and mortar" rather than professorial chairs because of the lure of a lasting monument. In wider use, the term "architecture" always means "unchanging deep structure."

It is an illusion. New usages persistently retire or reshape buildings. The old church is torn down, lovely as it is, because the parishioners have gone and no other use can be found for it. The old factory, the plainest of buildings, keeps being revived: first for a collection of light industries, then for artists' studios, then for offices (with boutiques and a restaurant on the ground floor), and something else is bound to follow. From the first drawings to the final demolition, buildings are shaped and reshaped by changing cultural currents, changing real-estate value, and changing usage.

The word "building" contains the double reality. It means both "the action of the verb BUILD" and "that which is built"—both verb and noun, both the action and the result. Whereas "architecture" may strive to be permanent, a "building" is always building and rebuilding. The idea is crystalline, the fact fluid. Could the idea be revised to match the fact?

That's the intent of this book. My approach is to examine buildings as a whole—not just whole in space, but whole in time. Some buildings are designed and managed as a spatial whole, none as a temporal whole. In the absence of theory or standard practice in the matter, we can begin by investigating: What happens anyway in buildings over time?

Two quotes are most often cited as emblems of the way to

David C. Fischetti.   Reprinted from *Old House Journal* (June 1991), p. 32.

1981 - THE TRUE NATURE OF BUILDINGS—that they can't hold still—is betrayed by a brick mansion on the move in Raleigh, North Carolina. The Capehart-Crocker house (1898) was moved to make room for a state government complex. The house is now used for offices.

understand how buildings and their use interact. The first, echoing the whole length of the 20th century, is "*Form ever follows function.*" Written in 1896 by Louis Sullivan, the Chicago highrise designer, it was the founding idea of Modernist architecture.[1] The very opposite concept is Winston Churchill's "*We shape our buildings, and afterwards our buildings shape us.*"[2] These were clairvoyant insights, pointing in the right direction, but they stopped short.

Sullivan's form-follows-function misled a century of architects into believing that they could really anticipate function. Churchill's ringing and-then-they-shape-us truncated the fuller cycle of reality. First we shape our buildings, then they shape us, then we shape them again—ad infinitum. Function reforms form, perpetually.

"Flow, continual flow, continual change, continual transformation" is how a Pueblo Indian architectural historian named Rina Swentzel describes her culture and her home village.[3] That describes everyone's culture and village.

In this century the houses of America and Europe have been altered utterly. When servants disappeared from them, kitchens suddenly grew, and servant's rooms became superfluous and

were rented out. Cars came, grew in size and number, then shrank in size, and garages and car parks tried to keep pace. "Family rooms" expanded around the television.   In the 1960s, women joined the work force, transforming both the workplace and the home. With shifting economic opportunities and stresses, families fragmented so much that the conventional nuclear family has become a rarity, and the design of housing is still catching up with that.

Office buildings are now the largest capital asset of developed nations and employ over half of their workforces. At the office, management theories come and go, each with a different physical layout. Unremitting revolutions in communication technology require rewiring of whole buildings every seven years on average. After the 1973 oil crisis, the energy budget of a building suddenly became a major issue, and windows, insulation, and heating and

---

1 Louis Sullivan, "The Tall Building Artistically Considered," *Lippincott's* (March 1896), pp. 403-409. This much-anthologized, beautifully bombastic essay climaxes with: "It is the pervading law of all things organic, and inorganic, of all things physical and metaphysical, of all things human and all things superhuman, of all true manifestations of the head, of the heart, of the soul, that the life is recognizable in its expressions, that form ever follows function." But when Sullivan applies the law to buildings, he adds a proviso that has been little noticed and never quoted: "Is it really then…so near a thing to us that we cannot perceive that the shape, form, outward expression, design or whatever we may choose, of the tall office building should in the very nature of things follow the functions of the building, and that where the function does not change, the form is not to change?" Mark that. "Where function does not change, form does not change." What about when function changes?

2 Churchill liked the statement so much he used it twice, first in 1924 to an awards ceremony for the Architectural Association, then before a national audience in 1943 on the occasion of requesting that the bomb-damaged Parliament be rebuilt exactly as it was before. To the architects he said, "There is no doubt whatever about the influence of architecture and structure upon human character and action. We make our buildings and afterwards they make us. They regulate the course of our lives." In Parliament, he restated it, "We shape our buildings, and afterwards our buildings shape us." Both times his example was the cramped, oblong Chamber of the House of Commons. It was to the good, he insisted, that the Chamber was too small to seat all the members (so great occasions were standing-room occasions), and that its shape forced members to sit on either one side or the other, unambiguously of one party or the other. "The party system, indeed, depends on the shape of the House of Commons," he concluded in 1924. [I am indebted to Marvin Nicely and Richard Langworth of the International Churchill Society for tracking down the quotes.]

3 Quoted by Jane Brown Gilette, "On Her Own Terms," *Historic Preservation* (Nov. 1992), p. 84.

October 1941. Library of Congress. Neg. no. LC-USF-81173-C.

1941 - RICH TO POOR? It looks at first glance like the prospects of this Coxsackie, New York, farm have gone downhill from left to right. I suspect that's why my old photography teacher, John Collier, took the photo for the Farm Security Administration—as an illustration of the harsh effects of the Depression. But building historian Dell Upton bets that the middle part was built first, in the 1820s. Then the fancy part was added on the left in the 1830s, and the kitchen moved into its own addition to the right in the 1850s.

BUILDINGS TELL STORIES, if they're allowed—
if their past is flaunted rather than concealed.

July 1972. Robert Nugent.

1972 - POOR TO RICH? No, stranger than that. After young Stephen W. Dorsey was elected US Senator from Arkansas in 1874, he made a pile of money with land and cattle speculation. In the remote northeast New Mexico prairie he constructed a mansion to suit his fortune and fame. It began in 1878 with frontier-romantic logs (left) and then shifted to Gothic-romantic sandstone (right) in 1881—complete with stone portraits on the upper tower of Dorsey and his wife and brother, plus two gargoyles in the likeness of his political enemy, Senator James Blaine. Dorsey got government contracts for mail delivery which later were investigated for fraud—$2 million had been stolen. The trial ruined Dorsey, and the mansion was foreclosed in 1893. Subsequent ranch families left the place as it was. It is now a house museum, a monument to frontier chicanery.

1990 - BLUE TO WHITE COLLAR. It was built as a valve factory in the 1930s in Emeryville, California. Now it houses 28 professional offices and live/work spaces—software designers, architects, photographers, and a magazine, *The Monthly*. When the factory at 1301 59th Street was gutted in 1985, a second floor was added throughout, providing 65,000 square feet total. Freight trains still rumble through several times a day, but the area has made the switch from dying-industrial to blooming-professional.

1992 - LIKE A MOUSE IN A COW SKULL, one specialty makes a home in another specialty's husk. Gas stations such as this one between the airport and the freeway in Albuquerque, New Mexico, are basically disposable buildings, left standing while the landlord waits for a big real estate score. Meanwhile, why not get some rent from the local karate club? It looks not bad as a dojo—lots of parking, and no neighbors to complain about the shouting.

May 1990. Brand.

9 March 1992. Brand.

cooling systems had to be completely revamped toward energy efficiency.

Asbestos went from being very good for you to very bad for you. Fire codes and building codes discovered new things to worry about, and old buildings were forced to meet the new standards. Access for the disabled transformed toilets, stairs, curbs, elevators.

Deterioration is constant, in new buildings as much as old. The roof leaks. The furnace is dying. The walls have cracks. The windows are a disgrace. People are getting sick from something in the air conditioning. The whole place is going to have to be redone!

And you can't fix or remodel an old place in the old way. Techniques and materials keep changing. Factory-hung windows and doors are better than the old site-built ones, but they have different shapes. Sheetrock replaces plaster; steel studs replace wood. You have to have vapor barriers, plastic plumbing, plastic electrical fixtures, a dozen new forms of insulation, track lighting, task lighting, uplighting, and carpet by the acre. The extent of change can be documented in *Architectural Graphic Standards*, the American builder's bible for design and construction details. It was first published in 1932. Selling in the hundreds of thousands, it was up to its eighth completely revised edition in 1988—with only part of one of its 864 pages still the same after 56 years. More than half of the 1988 edition was new or revised since the 1981 edition—seven routine years.

More is being spent on changing buildings than on building new ones. At the end of the 1980s, one of the new preservation professionals, Sally Oldham, could report formidable statistics. Home renovation in America had more than doubled during the decade. Commercial rehabilitation expenditures had gone from three-fourths of new construction to one-and-a-half times new construction. Some $200 billion (5 percent of the gross national product) was spent on renovation and rehabilitation in 1989, and historic preservation accounted for $40 billion a year of goods and services.[4] Nearly all architects (96 percent) were involved in some

form of rehabilitation, and a quarter of architects' revenues came from rehab.

Buildings keep being pushed around by three irresistible forces—technology, money, and fashion. Technology offers, say, new double-pane insulated windows with a sun-reflective membrane—expensive, but they will save enormously on energy costs for the building, and you get political points for installing them. By the time their defects become intolerable, even newer windows will beckon. The march of technology is inexorable, and accelerating.

Form follows funding. If people have money to spare, they *will* mess with their building, at minimum to solve the current set of frustrations with the place, at maximum to show off their wealth, on the reasonable theory that money attracts money. A building is not primarily a building; it is primarily property, and as such, subject to the whims of the market. Commerce drives all before it, especially in cities. Wherever land value is measured in square feet, buildings are as fungible as cash. Cities devour buildings.

As for fashion, it is change for its own sake—a constant unbalancing of the status quo, cruelest perhaps to buildings, which would prefer to remain just as they are, heavy and obdurate, a holdout against the times. Buildings are treated by fashion as big, difficult clothing, always lagging embarrassingly behind the mode of the day. This issue has nothing to do with function: fashion is described precisely as "non-functional stylistic dynamism" in *Man's Rage for Chaos* by Morse Peckham.[5] And fashion is culture-wide and inescapable.

---

4 Sally G. Oldham, "The Business of Preservation is Bullish and Diverse," *Preservation Forum*, National Trust for Historic Preservation (Winter, 1990), p. 14.

5 Morse Peckham, *Man's Rage for Chaos*, (New York: Chilton, 1965). Morse holds that the edge of fashion is art, and "art is the exposure to the tensions and problems of a false world so that man may endure exposing himself to the tensions and problems of the real world." We practice meaningless change in order to tolerate necessary change. That's fine, but in buildings the meaningless change of fashion often obstructs necessary change.

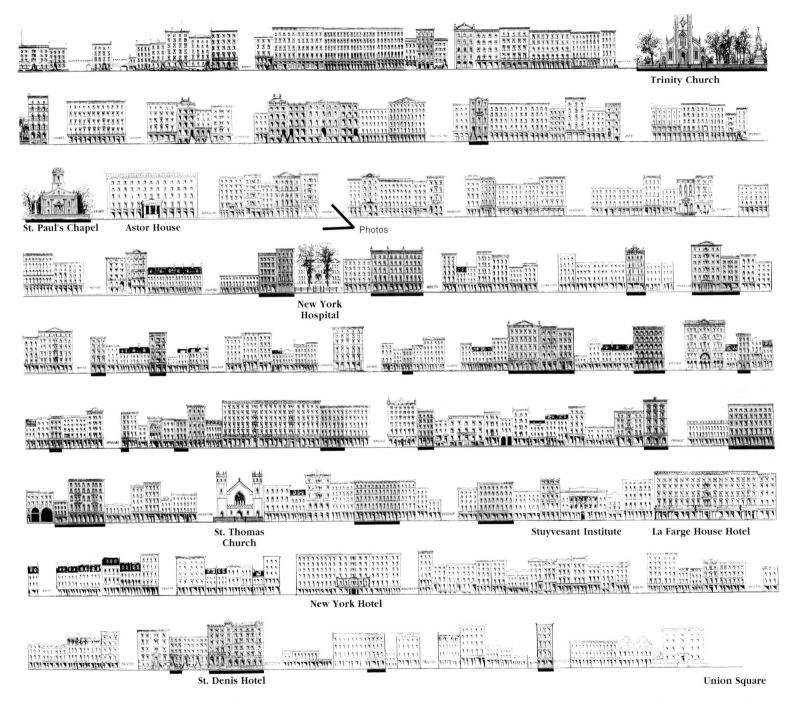

Trinity Church

St. Paul's Chapel   Astor House   Photos

New York
Hospital

St. Thomas
Church   Stuyvesant Institute   La Farge House Hotel

New York Hotel

St. Denis Hotel   Union Square

CITIES DEVOUR BUILDINGS. In 1865 the west side of lower Broadway in New York had 261 buildings, shown here. By 1990 only 33 were still there—one in eight. The survivors are shown shaded, with thickened underline.

Reprinted from David W. Dunlap's spectacular *On Broadway* (New York: Rizzoli, 1990), p. 4.

1880 - BROADWAY, west side, looking south from Park Place—the view matches the left half of the third row in the illustration to the left. On the corner of Broadway and Park Place is the Berkshire Life Insurance Building (1852). The wide white building beyond is the Astor House hotel (1836), and peeking out behind it is the portico of St. Paul's Chapel (1766).

3 December 1974. Edmund V. Gillon, Jr. Both photos and the caption information are from Edward B. Watson, *New York Then and Now* (New York: Dover, 1976), pp. 8-9. See Recommended Bibliography.

1974 - Same view as above. The entire block between Park Place and Barclay has become the world-famous neo-Gothic Woolworth Building (1913). Astor House was replaced by the Transportation Building (1927) and six-story Franklin Building (1914). Only St. Paul's Chapel remains from the previous photo. The United States Steel Building (1972) towers in the background.

To say that change in buildings is nearly universal does not help much in understanding how the process works, nor in conjuring how it might go better. Could different kinds of change be contrasted? Early in the research for this book, Sim Van Der Ryn, former State Architect of California, suggested I pay attention to the three different kinds of buildings, which he thought changed in quite separate ways—commercial, domestic, and institutional.

Commercial buildings have to adapt quickly, often radically, because of intense competitive pressure to perform, and they are subject to the rapid advances that occur in any industry. Most businesses either grow or fail. If they grow, they move; if they fail, they're gone. Turnover is a constant. Commercial buildings are forever metamorphic.

Domestic buildings—homes—are the steadiest changers, responding directly to the family's ideas and annoyances, growth and prospects. The house and its occupants mold to each other twenty-four hours a day, and the building accumulates the record of that intimacy. That is far less the case with renters, who must ask permission from landlords and have no hope of financial gain from improvements, but two-thirds of Americans (and Britons) own their homes.

Institutional buildings act as if they were designed specifically to prevent change for the organization inside and to convey timeless reliability to everyone outside. When forced to change anyway, as they always are, they do so with expensive reluctance and all possible delay. Institutional buildings are mortified by change.

The three kinds of buildings diverge from each other deliberately. The crass seething of commerce is something that institutional buildings seek to rise above and that homes seek to escape. But most institutional buildings are just offices after all, and offices are infamously high-change environments, and so they are self-violating. Domestic buildings are a successful sanctuary only when property values are constant, which is seldom.

Each kind of building also has different internal dynamics. Buildings whose business it is to make money signal when they are failing—annual cost exceeds income—and then usage and structure keep being adjusted until there's a fit (usually temporary). Institutional buildings house bureaucracies, which are not allowed to fail and so cannot help outgrowing their space.

8

1938 - A typical brick multi-use commercial building in Lawrence, Kansas, houses a movie theater, coffee shop, bus depot, and what might be offices upstairs on the far left. The style is 1930s Hollywood.

1979 - Forty years later, only the basic structure of the building and the theater's name are the same. The theater part appears to have been remodeled in the 1950s (probably made larger inside, since the previous windows and doors at the far left have disappeared). Lawrence, a college town, had its downtown upgraded in the 1970s. One result is the former bus depot being converted to suave professional offices.

COMMERCIAL AND DOMESTIC BUILDINGS CHANGE DIFFERENTLY—
commercial more kaleidoscopically (above), domestic more steadily (below).

ca. 1900 - BUILDINGS ALWAYS GROW. Even confined on a corner lot in San Francisco (at Hyde and Lombard), the Mrs. Robert Louis Stevenson-Lloyd Osbourne house contrived ways to grow.

ca. 1939 - Since 1900 the house has pushed out onto its upstairs deck. A garage has appeared on the left, and bay windows on the right. The tall brick chimneys no doubt were shaken down in the 1906 earthquake.

ca. 1870 - A mint was established in San Francisco to handle the millions of dollars of gold and silver flowing from the mines of the 49ers. This architect's drawing shows what was constructed in 1874 of Sierra granite and Columbia bluestone, with cast iron pillars and wrought iron girders. Despite the fact of having outgrown a previous mint building (1854), the officials and the architect made no provision in this building for later growth.

INSTITUTIONAL BUILDINGS DEFY CHANGE. The old US Mint in San Francisco has not changed since 1874, nor will it. It ignored an earthquake and shrugged off losing its function.

1906 - The 1906 earthquake and fire devastated the city around it, but iron security shutters and heroic efforts of staff protected the building and the $308 million in gold inside. The money was used to back banks that restored commerce and rebuilt San Francisco.

ca. 1941 - In 1940 the whole house surged upward one and two stories. Windows multiplied. The garage became three garages. Each change was an extension or increase of what was there before, rather than a transformation.

1992 - In 1937 the minting activity outgrew the building and moved into a new and larger structure (even more monumental) a mile away. The Old US Mint building, physically unchanged, still part of the government, lives on as mostly a museum of itself ("See $1 million in gold")—half the ground floor and basement. The Department of the Treasury has offices there to handle orders for commemorative coins and medals. And the massive chimneys still work for a living—the furnaces that once melted gold now heat many nearby buildings.

18 October 1990. Brand.

1990 - A classic single house is the Thomas Legare house (circa 1759) at 90 Church Street in Charleston. Of the 5,000 18th-century and 19th-century dwellings still standing in the city, some 3,000 are single houses. The characteristic piazza (double porch) was always built on the south or west side of the house—to give protection from the summer sun, access to the winter sun, and a place to enjoy Charleston's treasured sea breeze. The piazzas serve as outside hallways for the narrow buildings.

A FIX BECOMES A FEATURE. Add-ons often become a distinctive part of a generic building type. In early Charlestown, South Carolina, a double-story "piazza" (porch) was added on to the British-style townhouses to make them livable in the hot, humid climate. It soon became a famed vernacular—the Charleston "single house." Similarly, cast-iron balconies added on to New Orleans buildings (often to replace rotting wood balconies) became part of that city's character. Even flying buttresses on cathedrals were a fix that became a feature.

Turf battles become vicious; eventually some activities overflow awkwardly into nearby buildings.

Homes are the domain of slowly shifting fantasies and rapidly shifting needs. The widowed parent moves in; the teenager moves out; finances require letting out a room (new door and outside stair); accumulating stuff needs more storage (or public storage frees up some home space); a home office or studio becomes essential. Meanwhile, desires accumulate for a new deck, a hot tub, a modernized kitchen, a luxurious bathroom, a walk-in closet, a hobby refuge in the garage, a kid refuge in the basement or attic, a whole new master bedroom.

There is a universal rule—never acknowledged because its action is embarrassing or illegal. *All buildings grow.* Most grow even when they're not allowed to. Urban height limits and the party walls of row houses, for instance, are no barrier. The building will grow into the back yard and down into the ground—halfway under the street in parts of Paris.

A question I asked everyone while working on this book was "What makes a building come to be loved?" A thirteen-year-old boy in Maine had the most succinct answer. "Age," he said. Apparently the older a building gets, the more we have respect and affection for its evident maturity, for the accumulated human investment it shows, for the attractive patina it wears—muted bricks, worn stairs, colorfully stained roof, lush vines.

Age is so valued that in America it is far more often fake than real. In a pub-style bar and restaurant you find British antique oak wall paneling—perfectly replicated in high-density polyurethane. On the roof are fiber-cement shingles molded and colored to look like worn natural slate. But Europe has its own versions of fakery, now themselves respectable with age—the picturesque ersatz ruins of 18th-century landscape gardening, 19th-century buildings pretending to be medieval, neoclassical columns always bone-white instead of wearing the original Greek or Roman bright colors.

It seems there is an ideal degree of aging which is admired. Things should not be new, but neither should they be rotten with age (except in New Orleans, which fosters a cult of decay). Buildings should be just ripe—worn but still fully functional. Genuinely old buildings are constantly refreshed, but not too far, and new buildings are forced to ripen quickly. Hence the fashion in wood shingles, which weather handsomely in the course of a single winter. They are expensive and a fire hazard and will need replacing all too soon, but never mind.

The widespread fakery makes us respect honest aging all the more. The one garment in the world with the greatest and longest popularity—over a century now—is Levi's denim blue jeans. Along with their practical durability, they show age honestly and elegantly, as successive washings fade and shrink them to perfect fit and rich texture. Ingenious techniques to simulate aging of denim come and go, but the basic indigo 501s, copper-riveted, carry on for decades. This is highly evolved design. Are there blue-jeans buildings among us? How does design honestly honor time?

We admire the grand gesture in architecture, but we respect something else. In a computer teleconference on design, Brian Eno, the British rock musician and avant-garde artist, wrote:

> We are convinced by things that show internal complexity, that show the traces of an interesting evolution. Those signs tell us that we might be rewarded if we accord it our trust. An important aspect of design is the degree to which the object involves you in its own completion. Some work invites you into itself by not offering a finished, glossy, one-reading-only surface. This is what makes old buildings interesting to me. I think that humans have a taste for things that not only show that they have been through a process of evolution, but which also show they are still a part of one. They are not dead yet.

Between the dazzle of a new building and its eventual corpse, when it is either demolished or petrified for posterity as a museum, are the lost years—the unappreciated, undocumented, awkward-seeming time when it was alive to evolution. If Eno is right, those are the best years, the time when the building can engage us at our own level of complexity. How do those years work, actually?

This book attempts to answer that, or at least to frame the question in helpful detail. The argument goes as follows. Buildings are layered by different rates of change (Chapter 2—Shearing Layers). Adaptation is easiest in cheap buildings that no one cares about (3—The Low Road) and most refined in long-lasting sustained-purpose buildings (4—The High Road). Adaptation, however, is anathema to architects and to most of the building professions and trades (5—Magazine Architecture). And the gyrations of real-estate markets sever continuity in buildings (6—Unreal Estate). The building preservation movement arose in rebellion, deliberately frustrating creative architects and the free market in order to restore continuity (7—A Quiet, Populist, Conservative, Victorious Revolution). Focus on preservation brought a new focus on maintenance (8—The Romance of Maintenance), and respect for humble older buildings brought investigation of their design wisdom by vernacular building historians (9—How Buildings Learn from Each Other). The same kind of investigation can be made of the persistent change, mostly amateur, that occurs in contemporary houses and offices (10—Function Melts Form).

With that perspective backward in mind, it is possible to rethink perspective forward (11—The Scenario-Buffered Building) and to imagine designing buildings that invite adaptation (12—Built for Change). Doing it right requires an intellectual discipline that doesn't yet exist (Appendix—The Study of Buildings in Time). The study is worth undertaking because, more than any other human artifact, buildings excel at improving with time, if they are given the chance.

And they are wonderful to study. All dressed up in layers of dissimulation, buildings are so naked.

CHAPTER 2

Shearing Layers

HERE'S A PUZZLE. On most American magazine racks you'll find a slick monthly called *Architectural Digest.* Inside are furniture and decor ads and articles with titles like "Unstudied Spaces in Malibu" and "Paris, New York (20th-Century French Pieces Transform an East Side Apartment)." Almost no architecture. The magazine's subtitle reads: "The International Magazine of Fine Interior Design."

Architects and interior designers revile and battle each other. Interior design as a profession is not even taught in architecture departments. At the enormous University of California, Berkeley, with its prestigious Environmental Design departments and programs, architecture students can find no course on interior design anywhere. They could take a bus several miles to the California College of Arts and Crafts, which does teach interior design, but no one takes that bus.

How did *Architectural Digest* manage to jump the chasm? Advertisers, the market, and a profound peculiarity of buildings did it. Originally, back in 1920, it *was* an architecture magazine, though for a public rather than a strictly professional audience. Gradually the magazine noticed that its affluent readers rebuilt interiors much more often than they built houses. After 1960, the advertisers, followed dutifully by the editors, migrated away from exterior vision toward interior revision—toward decorous remodeling—where the action and the money were. The peculiarity of buildings that turned *Architectural Digest* into a contradiction of itself is that different parts of buildings change at different rates.

The leading theorist—practically the only theorist—of change rate

Cooper-Hewitt Museum, National Museum of Design, Smithsonian Institution/Art Resource, New York. Neg. no. 1960-211-1 A to M.

INTERIORS ARE FLIGHTY, fickle, and inconstant— whether from caprice, or wear and tear, or the irregular shifts of necessity.

Boredom plus money plus fashion equals new wallpaper every seven years. So it was in the Nathan Beers house of Fairfield, Connecticut. Thirteen consecutive layers of wallpaper were pasted over one another between the 1820s and 1910. This display is at the Cooper-Hewitt Museum in New York City.

in buildings is Frank Duffy, cofounder of a British design firm called DEGW (he's the "D"), and president of the Royal Institute of British Architects for 1993 to 1995. "Our basic argument is that there isn't such a thing as a building," says Duffy. "A building properly conceived is several layers of longevity of built components." He distinguishes four layers, which he calls Shell, Services, Scenery, and Set. Shell is the structure, which lasts the lifetime of the building (fifty years in Britain, closer to thirty-five in North America). Services are the cabling, plumbing, air conditioning, and elevators ("lifts"), which have to be replaced every fifteen years or so. Scenery is the layout of partitions, dropped ceilings, etc., which changes every five to seven years. Set is the shifting of furniture by the occupants, often a matter of months or weeks.

Like the advertisers of *Architectural Digest*, Duffy and his architectural partners steered their firm toward the action and the

---

1 Francis Duffy, "Measuring Building Performance," *Facilities* (May 1990), p. 17.

Over fifty years, the changes within a building cost three times more than the original building. Frank Duffy explains this diagram: "Add up what happens when capital is invested over a fifty-year period: the Structure expenditure is overwhelmed by the cumulative financial consequences of three generations of Services and ten generations of Space plan changes. That's the map of money in the life of a building. It proves that architecture is actually of very little significance—it's nugatory."[1] (I have translated Duffy's terms into my terms.)

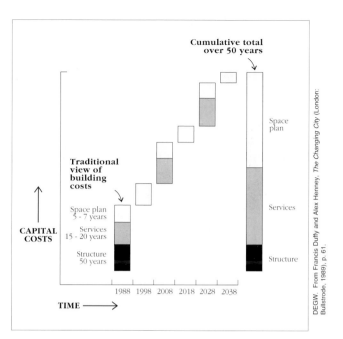

DEGW. From Francis Duffy and Alex Henney, *The Changing City* (London: Bullstrode, 1989), p. 61.

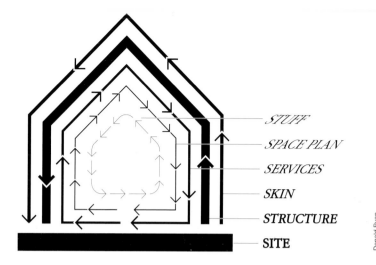

Donald Ryan

SHEARING LAYERS OF CHANGE. Because of the different rates of change of its components, a building is always tearing itself apart.

money. DEGW helps rethink and reshape work environments for corporate offices, these days with a global clientele. "We try to have long-term relationships with clients," Duffy says. "The unit of analysis for us isn't the building, it's the use of the building through time. Time is the essence of the real design problem."

I've taken the liberty of expanding Duffy's "four S's"—which are oriented toward interior work in commercial buildings—into a slightly revised, general-purpose "six S's":

- SITE - This is the geographical setting, the urban location, and the legally defined lot, whose boundaries and context outlast generations of ephemeral buildings. "Site is eternal," Duffy agrees.

- STRUCTURE - The foundation and load-bearing elements are perilous and expensive to change, so people don't. These *are* the building. Structural life ranges from 30 to 300 years (but few buildings make it past 60, for other reasons).

- SKIN - Exterior surfaces now change every 20 years or

so, to keep up with fashion or technology, or for wholesale repair. Recent focus on energy costs has led to re-engineered Skins that are air-tight and better-insulated.

- SERVICES - These are the working guts of a building: communications wiring, electrical wiring, plumbing, sprinkler system, HVAC (heating, ventilating, and air conditioning), and moving parts like elevators and escalators. They wear out or obsolesce every 7 to 15 years. Many buildings are demolished early if their outdated systems are too deeply embedded to replace easily.

- SPACE PLAN - The interior layout—where walls, ceilings, floors, and doors go. Turbulent commercial space can change every 3 years or so; exceptionally quiet homes might wait 30 years.

- STUFF - Chairs, desks, phones, pictures; kitchen appliances, lamps, hair brushes; all the things that twitch around daily to monthly. Furniture is called *mobilia* in Italian for good reason.

1863 - The first Cliff House was built in 1863 to take commercial advantage of the site's spectacular view of Seal Rocks (which crawled with sea lions) and Pacific sunsets. The restaurant's commercial success was always tenuous, because its customers in San Francisco were seven sandy miles away to the east.

1878 - In 1868 the original owner tripled the Cliff House in size with two asymmetric wings and a long roofed balcony. It was by now a successful gambling casino. San Francisco silver-mining millionaire Adolph Sutro, having built a home and public garden on the heights overlooking the Cliff House, didn't like its rowdy reputation, so he bought the place and converted it to a family restaurant. On Christmas night in 1894 a kitchen fire burned the building down.

(NATURAL) SITE IS ETERNAL. At San Francisco's famous Cliff House, the house

ca. 1910 - Sutro's daughter, Dr. Emma Merritt, had the next Cliff House made of fireproof concrete and steel, designed by brothers named Reid. President Taft dined at this one.

ca. 1946 - After a series of owners and a number of years being closed, the Cliff House was bought and refurbished by George and Leo Whitney in 1937. It featured the largest curio shop in the world.

Marilyn Blaisdell Collection.

William C. Billington. Marilyn Blaisdell Collection.

Charles T. Hall. Marilyn Blaisdell Collection.

Turrill & Miller. Marilyn Blaisdell Collection.

1895 - Adolph Sutro was an engineer accustomed to huge projects. He hired architects C. J. Colley and Emile S. Lemme to design a chateau-style edifice to match the grandeur of the site.

ca. 1900 - Presidents McKinley and Roosevelt dined at Sutro's Cliff House. Eight stories high, it had art galleries and ballrooms as well as dining rooms and bars. Sutro began building a railroad to bring customers to his amusement palace. Solidly nailed to its cliff with iron rods, the building suffered no damage at all from the great earthquake and fire of 1906.

1907 - On September 7, 1907, the dream burned to rubble, with just a few chimneys left standing.

comes and goes. The cliff stays.

1973 - The Cliff House closed again in 1969, then reopened in 1973—during San Francisco's "psychedelic" heyday—with a heady mural of ocean waves and spray. Most of the photos and information on these two pages are from *San Francisciana: Photographs of the Cliff House* (San Francisco: Blaisdell, 1985), by Marilyn Blaisdell. See Recommended Bibliography.

1991 - In 1977 the National Park Service took over the Cliff House as part of the Golden Gate National Recreation Area. It is appropriately staid and public-spirited in demeanor. Every so often someone revives the fantasy of rebuilding Sutro's extravaganza. Stranger things have happened.

California Historical Society, San Francisco. Neg. no. FN-27351.

Gregory Gaar. Marilyn Blaisdell Collection.

25 October 1991. Brand.

ca. 1954 - In 1950 the Whitney brothers drastically remodeled the building with redwood siding and extended it to the left. The building never did get its dignity back.

16

1860 - Looking due east across what is now the financial district of Boston, this was the first aerial photo of an American city—shot by J. W. Black from a balloon tethered at 1,200 feet. Keep your eye on the steepled church at the far left, the Old South Meeting House. This pair of photos is reprinted from the excellent rephotography book, *Cityscapes of Boston*, by Robert Campbell and Peter Vanderwarker (Boston: Houghton Mifflin, 1992. See Recommended Bibliography.)

1981 - Twelve decades later every single building but one—the Old South Meeting House at far left—is gone. What the great fire of 1872 did not take, real estate pressures did. But the streets are entirely intact, and buildings like the parking lot bent in the middle foreground and the Shawmut Bank building tall and trapezoidal in the middle top, must twist to fit the streets and their angular lots. Milk Street is the curving street on the left; Franklin curves on the right. Washington Street angles down in the foreground from the Old South Meeting House.

**(POLITICAL) SITE IS ETERNAL. The streets of Boston, tangled as they are, won't move. Even the skyscrapers must dance to their choreography.
In Lucca, Italy, the outline of a Roman amphitheater lives on in the modern city.**

ca. 1980 - The oval of an ancient Roman amphitheater in Lucca, Italy, was preserved by gradually turning into private property. When the empire died and the entertainment stopped, people moved into the obsolete structure and made their homes and shops there. Over the centuries all of the original structure was replaced, but its outline persisted in the property lines. The center of the oval eventually became crowded with buildings. The space was reopened into a piazza in the 19th century, the better to attract tourists to the ghost amphitheater.

Duffy's time-layered perspective is fundamental to understanding how buildings actually behave. The 6-S sequence is precisely followed in both design and construction. As the architect proceeds from drawing to drawing—layer after layer of tracing paper—"What stays fixed in the drawings will stay fixed in the building over time," says architect Peter Calthorpe. "The column grid will be in the bottom layer." Likewise the construction sequence is strictly in order: Site preparation, then foundation and framing the Structure, followed by Skin to keep out the weather, installation of Services, and finally Space plan. Then the tenants truck in their Stuff.

Frank Duffy: "Thinking about buildings in this time-laden way is very practical. As a designer you avoid such classic mistakes as solving a five-minute problem with a fifty-year solution, or vice versa. It legitimizes the existence of different design skills—architects, service engineers, space planners, interior designers—all with their different agendas defined by this time scale. It means you invent building forms which are very adaptive."

The layering also defines how a building relates to people. Organizational levels of responsibility match the pace levels. The building interacts with individuals at the level of Stuff; with the tenant organization (or family) at the Space plan level; with the landlord via the Services (and slower levels) which must be maintained; with the public via the Skin and entry; and with the whole community through city or county decisions about the footprint and volume of the Structure and restrictions on the Site. The community does not tell you where to put your desk or your bed; you do not tell the community where the building will go on the Site (unless you're way out in the country).

Buildings rule us via their time layering at least as much as we rule them, and in a surprising way. This idea comes from Robert V.

O'Neill's *A Hierarchical Concept of Ecosystems*. O'Neill and his co-authors noted that ecosystems could be better understood by observing the rates of change of different components. Hummingbirds and flowers are quick, redwood trees slow, and whole redwood forests even slower. Most interaction is within the same pace level—hummingbirds and flowers pay attention to each other, oblivious to redwoods, who are oblivious to them. Meanwhile the forest is attentive to climate change but not to the hasty fate of individual trees. The insight is this: "*The dynamics of the system will be dominated by the slow components, with the rapid components simply following along.*"[2] Slow constrains quick; slow controls quick.

The same goes with buildings: the lethargic slow parts are in charge, not the dazzling rapid ones. Site dominates the Structure, which dominates the Skin, which dominates the Services, which dominate the Space plan, which dominates the Stuff. How a room is heated depends on how it relates to the heating and cooling Services, which depend on the energy efficiency of the Skin, which depends on the constraints of the Structure. You could add a seventh "S"—human Souls at the very end of the hierarchy, servants to our Stuff.

Still, influence does percolate the other direction. The slower processes of a building gradually integrate trends of rapid change within them. The speedy components propose, and the slow dispose. If an office keeps replacing its electronic Stuff often enough, finally management will insist that the Space plan acquire a raised floor to make the constant recabling easier, and that's when the air conditioning and electrical Services will be revamped to handle the higher load. Ecologist Buzz Holling points out that it is at the times of major changes in a system that the quick processes can most influence the slow.

The quick processes provide originality and challenge, the slow provide continuity and constraint. Buildings steady us, which we can probably use. But if we let our buildings come to a full stop, they stop us. It happened in command economies such as Eastern

---

2   R. V. O'Neill, D. L. DeAngelis, J.B Wade, T. F. H. Allen, *A Hierarchical Concept of Ecosystems* (Princeton: Princeton University Press, 1986), p. 98.

Europe's in the period 1945-1990. Since all buildings were state-owned, they were never maintained or altered by the tenants, who had no stake in them, and culture and the economy were paralyzed for decades.

Slow is healthy. Much of the wholesome evolution of cities can be explained by the steadfast persistence of Site. Property lines and thoroughfares in cities are inviolate even when hills are leveled and waterfronts filled in. After the Great Fire of London in 1666, the city was rebuilt of brick, with widened streets but upon the old ground plan, and with meticulously preserved property lines. A wise move, says urban scholar Kevin Lynch: "Rebuilding was rapid and vigorous because each man could start again on his own familiar land."[3] Exactly the same thing happened two-and-a-half centuries later in San Francisco, after its earthquake and fire of 1906.

Different Site arrangements lead to different city evolutions. Downtown New York City, with its very narrow long blocks, is uniquely dense and uniquely flexible. Quick-built San Francisco

is kept adaptable, congenial, and conservative over the decades by its modest lot sizes, according to urban designer Anne Vernez Moudon:

> Small lots will support resilience because they allow many people to attend directly to their needs by designing, building, and maintaining their own environment. By ensuring that property remains in many hands, small lots bring important results: many people make many different decisions, thereby ensuring variety in the resulting environment. And many property owners slow down the rate of change by making large-scale real estate transactions difficult.[4]

After Site comes Structure, at the base of which is the all-determining foundation. If it is out-of-square or out-of-level, it will plague the builders clear to the roof line and bother

STRUCTURE PERSISTS AND DOMINATES. In Santa Fe's old State Capitol building, the original Structure defined the remodeling possibilites—even with radical changes of Skin, footprint, volume, and interior design.

ca. 1916 - Watch the two arched windows at the upper left of the facade. On this site the first New Mexico State Capitol building was built in 1886 and burned in 1892 (arson suspected.) This domed capitol went up in 1900 but soon proved embarrassing, because around 1920, Santa Fe decided to re-model the whole city to look histori-cal—Spanish Colonial and Territorial style (see Chapter 9 for the strange story).

Anna L. Hase. Museum of New Mexico. Neg. no. 16711.

1992 - The two arched windows are still there, and so is the core Structure, but not much else is visibly the same. Between 1950 and 1953 the building had its dome decapitated, its classical portico demolished, a tower added (along with extensive new space), and the whole thing tricked out with a particularly unconvincing adobe look and Territorial-style detailing such as the brick wall-tops. In 1966 a new Capitol building was built nearby, and this became the Bataan Memorial Building, still housing some state offices.

February 1992. Brand.

ca. 1865 - The US Soldiers' Home (1857) was originally designed to look Italianate by a second lieutenant in the US Army Corps of Engineers. The material was New York marble.

ca. 1872 - In 1868 the building camouflaged its expansion upward with a fashionable Second Empire mansard roof. The tower also grew higher and acquired a water tank.

ca. 1910 - In 1884 and 1887 the rear of the building was expanded in a Gothic Revival style, and in 1890 the front of the building caught up, growing still higher in the process. The building has housed veterans from every American war since the War of 1812.

SKIN IS MUTABLE. Even institutional buildings like the Soldiers' Home in Washington DC can't resist periodically shedding old skin for a new.

remodelers for the life of the building. If it is weak, it permanently limits the height of the building. If it lets in water or offers inadequate headroom for the basement, remedy is nearly impossible.

The mutability of Skin seems to be accelerating. Demographer Joel Garreau[5] says that in "edge cities" (new office and commercial developments on the periphery of older cities) developers are accustomed to fine-tune their buildings by changing rugs and facades—a typical "facadectomy" might go upscale from pretentious marble veneer to dignified granite veneer to attract a richer tenant. Developers expect their building Skins to "ugly out" every fifteen years or so, and plan accordingly.

The longevity of buildings is often determined by how well they can absorb new Services technology. Otis Elevator contractors don't bother to make money on their first installation. They know you'll be back soon enough for improved elevators; their profits are in the inevitable renovations. Energy Services such as electricity and gas are driven constantly toward greater efficiency by their sheer expense—30 percent of operating costs, equal over a building's life to the entire original cost of construction. Between the Energy Crisis of 1973 and 1990, the money spent on space heating in new American buildings dropped by a dramatic 50 percent.[6]

Even the home is no refuge from turnover in Services. Houses were revolutionized by the arrival of public water service around 1900, then by public electricity in the 1920s and 1930s, later by cable television in the 1970s. The two most renovated rooms in all houses are the kitchen and bathroom. Building historian Orlando Ridout says that in Maryland, you can find more whole houses from the 1700s than pre-1920 toilets. Whether it's the arrival of colored enamel in the 1920s, the advent of Jacuzzi baths in the 1970s, or guilt about water-wasting toilets in the 1980s, people keep making changes and expanding the significance of the bathroom in their homes. Likewise the kitchen, which has migrated from a back corner to the middle of home life, while stoves, refrigerators, and sinks are replaced as frequently as

[3] Kevin Lynch, *What Time Is This Place?* (Cambridge: MIT Press, 1972), p. 8. See Recommended Bibliography.

[4] Anne Vernez Moudon, *Built for Change* (Cambridge: MIT Press, 1986), p. 188. See Recommended Bibliography.

[5] Garreau is the author of *Edge City* (New York: Doubleday, 1991). See Recommended Bibliography.

[6] Rick Bevington and Arthur H. Rosenfeld, "Energy for Buildings and Homes," *Scientific American* (Sept. 1990), p. 77.

automobiles. Service-connected Stuff will not hold still.

The Space plan and Stuff are what building users have to look at and deal with all day, and they rapidly grow bored, frustrated, or embarrassed by what they see. Between constant tinkering and wholesale renovation, few interiors stay the same for even ten years.

A design imperative emerges: *An adaptive building has to allow slippage between the differently-paced systems of Site, Structure, Skin, Services, Space plan, and Stuff.* Otherwise the slow systems block the flow of the quick ones, and the quick ones tear up the slow ones with their constant change. Embedding the systems together may look efficient at first, but over time it is the opposite, and destructive as well.

Thus, pouring concrete on the ground for an instant foundation ("slab-on-grade") is maladaptive—pipes are foolishly buried, and there's no basement space for storage, expansion, and maintenance and Services access. Timber-frame buildings, on the other hand, conveniently separate Structure, Skin, and Services, while balloon-frame (standard stud construction) over-connects them.

All these shearing layers of change add up to a whole for the building, but how do they add up to a whole for the occupants? How can they change *toward* the humans in them rather than

T. Harmon Parkhurst. Museum of New Mexico. Neg. no. 50967.

ca. 1935 - A series of cafes have occupied this building on the corner of Don Gaspar and Water streets in Santa Fe, New Mexico. Its corner location near the central plaza kept it busy, but cafes are ephemeral enterprises.

T. Harmon Parkhurst. Museum of New Mexico. Neg. no. 50968.

ca. 1935 - A soda fountain and booths dominated the Space plan of the K. C. Waffle house. It said, just as clearly as the sign outside, "Tourists, come in as you are." Southwestern style is evident in the tile and leather.

L. C. Durette. Library of Congress. Neg. no. HABS NH-8-PORT, 124-15.

1936 - SERVICES OBSOLESCE AND WEAR OUT. In the kitchen of the Captain Barnes house (1808) in Portsmouth, New Hampshire, Services-connected appliances were layered on each other. Originally it had a large fireplace. About 1816 a contrivance called the Rumford Roaster was added on the left (round plate). Then a stove was built into the fireplace (probably 1840s), and a later stove (probably around 1900) crowded in front of it. Also visible are a water heater (cylinder behind the stove), rack for drying clothes on the right, and a bare electric light.

1991 - The K. C. Waffle House became the Mayflower Cafe, then Golden Temple Conscious Cookery (1974-1977), then Pogo's Eatery (1977-1979), then Cafe Pasqual's (1978-?).

1991 - The Space plan of Pasqual's features a raised seating area by the entrance and added rest rooms at the back. While the exterior shows a modest effort at deepening "authenticity," the interior motif is clamorously Mexican. The cooking is Santa Fe chic.

INTERIORS CHANGE RADICALLY while exteriors maintain continuity. The Space plan is the stage of the human comedy. New scene, new set.

away, as so many seem to do? Here the leading theorist is Christopher Alexander. A long-time professor of architecture at the University of California, Berkeley, Chris Alexander is the author of an influential series of books from Oxford University Press which explore in practical detail what it is that makes buildings and communities humane—or more precisely, what makes them become humane over time.[7]

A design professional of depth—his 1964 *Notes on the Synthesis of Form* is still in print—Alexander is inspired by how design occurs in the natural world. "Things that are good have a certain kind of structure," he told me. "You can't get that structure except dynamically. Period. In nature you've got continuous very-small-feedback-loop adaptation going on, which is why things get to be harmonious. That's why they have the qualities that we value. If it wasn't for the time dimension, it wouldn't happen. Yet here *we* are playing the major role in creating the world, and we haven't figured this out. That is a very serious matter."

Applying this approach to buildings, Alexander frames the design question so: "What does it take to build something so that it's really easy to make comfortable little modifications in a way that once you've made them, they feel integral with the nature and structure of what is already there? You want to be able to mess around with it and progressively change it to bring it into an

---

7 Alexander's "yellow books" from Oxford University Press, each with a variety of co-authors, are: *The Timeless Way of Building* (1979); *A Pattern Language* (1977); *The Oregon Experiment* (1975); *The Production of Houses* (1985); *The Linz Café* (1981); *A New Theory of Urban Design* (1987). See Recommended Bibliography. A reviewer in *Architectural Design* called *A Pattern Language* "perhaps the most important book on architectural design published in this century."

STUFF JUST KEEPS MOVING. The Treaty Room of the White House has had the same Space plan since 1817—except for a temporary partition installed for Abraham Lincoln in 1861. But the furniture and fittings blinked in and out of the room as administrations and fashions came and went, and the room's use varied from bedroom to outer office, to Cabinet room, to inner office, to sitting room, to library.

ca. 1911 - The White House was drastically remodeled by architect Charles McKim in 1902, during Theodore Roosevelt's administration. President William Taft (1909-1913) continued Roosevelt's use of the Treaty Room as his office. The door trim was the same, but the fireplace, ceiling cornice, furniture, shelving, rug, and pictures all were different.

ca. 1891 - On the second floor of the White House, what is now called the Treaty Room is connected by an inner door to the Oval Office. It has always been an intimate part of the President's family or work life. In President Benjamin Harrison's administration (1889-1893) it was used as the Cabinet Room, dominated by President Grant's table for Cabinet meetings in the middle. Watch the chairs, the rug, the chandelier, fireplace and its mirror, and the pictures on the wall.

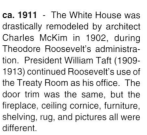

1931 - President Herbert Hoover (1929-1933) had a wife, Lou, whose ambition was to convert the Treaty Room to the "Monroe Drawing Room." It had been used as an inner sanctum office by President Woodrow Wilson, then as a sitting room by Warren Harding and Calvin Coolidge. Mrs. Hoover sought antique furniture from the period of President James Monroe (1817-1825). She found a chandelier that had been discarded from the Green Room (the state room just below this one).

ca. 1895 - President Grover Cleveland (1893-1897) has a different chair for himself at the head of the Cabinet table, and there's a new chandelier. A framed picture in the corner has changed, and the bookshelves there are gone. (In 1993 the Cabinet table was still in this room—as President William Clinton's desk.)

1961 - President John F. Kennedy (1961-1963) also had a wife, Jacqueline, who loved to supervise interior design. The room looked like this after Roosevelt, Truman, and Eisenhower. (During Truman's administration, the entire building was gutted and rebuilt—this room had the same trim, fireplace, and chandelier, but the walls, ceiling, floor, and windows were new construction. The Space plan had altered not an inch.)

ca. 1899 - President William McKinley (1897-1901) has a new chair, new rug, and a new fireplace screen, but the chandelier and the framed picture and bookshelves in the corner from Harrison's era have returned. McKinley was the first of several presidents to use the room as his private office. In this room he signed the declaration of war with Spain in 1898 and five months later signed with the ambassador of France the protocols for the peace conference—hence, the Treaty Room.

1963 - Jackie Kennedy helped found the White House Historical Association, and she brought in French interior designer Stefan Boudin. Their version of the Treaty Room restored over the fireplace a massive ornate mirror of the kind that was there when McKinley signed the peace protocols with Spain. On the wall is an 1899 painting of the very event. The table and chairs were kept, along with the chandelier that Mrs. Hoover fancied.

adapted state with yourself, your family, the climate, whatever. This kind of adaptation is a continuous process of gradually taking care." You can recognize the result where that process is working, he writes. "Because the adaptation is detailed and profound, each place takes on a unique character. Slowly, the variety of places and buildings begins to reflect the variety of human situations in the town. This is what makes the town alive."[8]

While all buildings change with time, only some buildings improve. What makes the difference between a building that gets steadily better and one that gets steadily worse? Growth, apparently, is independent of adaptation, and spasmodic occupant-turnover can defeat adaptation.

Growth follows a simple goal of property owners: maximize what you control. The practice is ancient. In old cities of Europe and the Mideast, upper stories would jetty out farther and farther to increase the space on each floor, until neighbors could shake hands across the street from upper rooms. Now as then, more space in domestic buildings is equated with freedom. In commercial buildings, more space means profit. In institutional buildings, it means power. Everyone tries to get more than they're allowed. City councils often seem to discuss little else. But only sometimes are additions an improvement. Adding more rooms around the periphery of a building, for instance, often leaves the middle dark and desolate.

The opposite of adaptation in buildings is graceless turnover. The usual pattern is for a rapid succession of tenants, each scooping out all trace of the former tenants and leaving nothing that successors can use. Finally no tenant replaces the last one, vandals do their quick work, and broken windows beg for demolition. There are two forms of surcease. If there is a turnaround in local real estate, the succession of owners and tenants might head back upscale, each one adding value. Or the building may be blessed with durable construction and resilient design which can forgive insult and hard swerves of usage. A brick factory from the 1910s, with its intelligent daylighting and abundant space, can stand empty for a decade and still gain value.

Age *plus adaptivity* is what makes a building come to be loved. The building learns from its occupants, and they learn from it.

There is precedent for thinking this way. In classical Greece and Rome, *domus* meant "house" in an expanded sense:

> People and their dwellings were indistinguishable: domus referred not only to the walls but also to the people within them. Evidence for this is found in inscriptions and texts, in which the word refers now to one, now to the other, but most often to both at once, to the house and its residents envisioned as an indivisible whole. The architectural setting was not an inert vessel; the genius of the domus, honored by a cult, was the protector of both the place and the people who lived in it.[9]

That kind of bonding between building and inhabitants still occurs. The next two chapters explore seemingly opposite examples of it—two kinds of buildings that easily become loved. One, grand and deep, I call the High Road—durable, independent buildings that steadily accumulate experience and become in time wiser and more respected than their inhabitants. The other, quick and dirty, is the Low Road. Their specialty is swift responsiveness to their occupants. They are unrespectable, mercurial, street-smart.

Among buildings as within them, differences of pace are everything.

---

8   Christopher Alexander, *The Timeless Way of Building* (New York: Oxford University Press, 1979), p. 231.

9   Yvon Thébert, "Private Life and Domestic Architecture in Roman Africa," *A History of Private Life*, 5 vols. (Cambridge: Harvard Univ. Press, 1985, 1987), vol. 1, p. 407.

CHAPTER 3

"Nobody Cares What You Do In There": The Low Road

IT HAS TO DO with freedom. Or so I surmised from a 1990 conversation with John Sculley, then head of Apple Computer. Sculley was trained in architecture before he started rocketing up corporate ladders. During a break at a conference, we got talking about buildings. Apple had expanded from five buildings into thirty-one in the few years Sculley had been at Apple. I asked him, "Do you prefer moving into old buildings or making new ones?" "Oh, old ones," he said. "They are much more freeing."

That statement throws a world of design assumptions upside down. *Why* are old buildings more freeing? A way to pursue the question is to ask, what kinds of old buildings are the most freeing?

A young couple moves into an old farmhouse or old barn, lit up with adventure. An entrepreneur opens shop in an echoing warehouse, an artist takes over a drafty loft in the bad part of town, and they feel joy at the prospect. They can't wait to have at the space and put it immediately to work. What these buildings have in common is that they are shabby and spacious. Any change is likely to be an improvement. They are discarded buildings, fairly free of concern from landlord or authorities: "Do what you want. The place can't get much worse anyway. It's just too much trouble to tear down."

Low Road buildings are low-visibility, low-rent, no-style, high-turnover. Most of the world's work is done in Low Road

buildings, and even in rich societies the most inventive creativity, especially youthful creativity, will be found in Low Road buildings taking full advantage of the license to try things.

Take MIT—the Massachusetts Institute of Technology. A university campus is ideal for comparing building effectiveness because you have a wide variety of buildings serving a limited number of uses—dormitories, laboratories, classrooms, and offices, that's about it. I'm familiar enough with MIT to know which two buildings are regarded with the most affection among the sixty-eight on campus. One, not surprisingly, is a dormitory called Baker House, designed by Alvar Aalto in 1949. Though Modernist and famous, it is warmly convivial and varied throughout, with a sintered-brick exterior that keeps improving with time.

But the most loved and legendary building of all at MIT is a surprise: a temporary building left over from World War II without even a name, only a number: Building 20. It is a sprawling 250,000-square-foot[1] three-story wood structure—"The only building on campus you can cut with a saw," says an admirer—constructed hastily in 1943 for the urgent development of radar and almost immediately slated for demolition. When I last saw it in 1993, it was still in use and still slated for demolition. In 1978 The MIT Museum assembled an exhibit to honor the perpetual fruitfulness of Building 20. The press release read:

> Unusual flexibility made the building ideal for laboratory and experimental space. Made to support heavy loads and of wood construction, it allowed a use of space which accommodated the enlargement of the working environment

---

[1] Roughly 25,000 square meters, if you use the rule of thumb: 10 square feet almost equals 1 square meter (0.929 square meters, to be exact).

ca. 1955 - Jules Barsotti's garage at 770 North Point in San Francisco had endearingly succinct ornament to go with its wide door and factory windows. Every building should feature its date of origin this way. A bas relief and spiffy roof trim is gravy.

1990 - A perfect general-purpose space, the old auto shop was an upscale retail outlet for Patagonia in the 1980s and 1990s, peddling high quality outdoor apparel.

1925 - It was no doubt sometimes too cold and sometimes too hot, but the big clear-span space of the garage had the early-20th-century daylighting that is more appreciated now that it is rare. The metal roof was cheap and effective and still sounds wonderful in the rain.

1993 - The 1924 Structure and Skin of the building was so simple and separated from everything else that it remains unchanged seventy years and many tenants later. With its abundant windows and airy steel truss overhead, it makes a cheery shop.

MADE STRICTLY FOR CARS, Barsotti's Automotive Service shop (1924) in San Francisco employed steel trusses to make its 3,000-square-foot space column-free so cars could maneuver easily. Years later, a florist, an aerobics studio, and then a clothing retailer found the space easily converted to their special needs. The building was too humble for anyone to worry about whether they were violating its historical or aesthetic integrity.

1945 - Here photographed from a Navy blimp at the end of World War II, the so-called Radiation Laboratory at Building 20 was one of its unsung heroes. In an undertaking similar in scope to the Manhattan Project that created the atomic bomb, the emergency development of radar employed the nation's best physicists in an intense collaboration that changed the nature of science. Unlike Los Alamos, the MIT radar project was not run by the military, and unlike Los Alamos, no secrets got out. The verdict of scientists afterward was, "The atom bomb only ended the war. Radar won it."

The MIT Museum. Neg. no. CC-20-417.

MIT'S LEGENDARY BUILDING 20 (1943) was an artifact of wartime haste. Designed in an afternoon by MIT grad Don Whiston, it was ready for occupancy by radar researchers six months later. With steel unavailable because of the war, the building was framed with heavy wood timbers. To do that required the city of Cambridge exempting the building from the fire code on the condition that it was a temporary structure. It is one of the strongest buildings on campus, capable of bearing 150 pounds per square foot.

The MIT Museum. Neg. no. CC-20-421.

1945 - During the war the innocuous building at 18 Vassar Street in Cambridge sprouted odd outgrowths overnight.

1945 - Its layout into five low, narrow wings gave good daylight throughout Building 20 and also provided courtyards between the wings that could be grown into when projects burst through their walls. One veteran of the building, Henry Zimmerman, commented, "I believe the horizontal layout helped to encourage interaction between groups. In a vertical layout with small floor areas, there is less research variety on each floor. Chance meetings in an elevator tend to terminate in the lobby, whereas chance meetings in a corridor tended to lead to technical discussions."

The MIT Museum. Neg. no. CC-20-410B.

19 August 1990. Brand.

1990 - Author Fred Hapgood wrote in 1993 of Building 20, "The edifice is so ugly that it is impossible not to admire it, if that makes sense; it has ten times the righteous nerdly swagger of any other building on campus." (*Up the Infinite Corridor*, New York: Addison-Wesley)

1990 - More discreet than during the war, Building 20's people now keep projects out of the courtyards and let semi-random plants grow there. The city of Cambridge is still trying to get the building demolished, in part because of the timber construction, in part because of the asbestos shingle siding. The occupants promise that they will not eat the shingles, which otherwise, they say, can cause no harm.

19 August 1990. Brand.

either horizontally or vertically. Even the roof was used for short-term structures to house equipment and test instruments.

Although Building 20 was built with the intention to tear it down after the end of World War II, it has remained these thirty-five years providing a special function and acquiring its own history and anecdotes. Not assigned to any one school, department, or center, it seems to always have had space for the beginning project, the graduate student's experiment, the interdisciplinary research center.

20 November 1990. Brand.

1990 - No one could accuse Building 20 of burying its Services too deep in the Structure. Recabling from office to office, lab to lab, or even wing to wing is largely a matter of do-it-yourself. Rather than a burden, the occupants consider this a benefit.

Indeed, MIT's first interdisciplinary laboratory, the renowned Research Laboratory of Electronics, founded much of modern communications science there right after the war. The science of linguistics was largely started there, and forty years later in 1993 one of its pioneers, Noam Chomsky, was still rooted there. Innovative labs for the study of nuclear science, cosmic rays, dynamic analysis and control, acoustics, and food technology were born there. Harold Edgerton developed stroboscopic photography there. New-technology companies such as Digital Equipment Corporation and Bolt, Baranek, and Newman incubated in Building 20 and later took its informal ways with them into their corporate cultures and headquarters. The Tech Model Railroad Club on the third floor, E Wing, was the source in the early 1960s of most of the first generation of computer "hackers," who set in motion a series of computer technology revolutions (still in progress).

1990 - The wide wood stairs in Building 20 show wear in a way that adds to its myth. You feel yourself walking in historic footsteps in pursuit of technical solutions that might be elegant precisely because they are quick and dirty. And that describes the building: elegant because it is quick and dirty.

20 November 1990. Brand.

Like most Low Road buildings, Building 20 was too hot in the summer, too cold in the winter, Spartan in its amenities, often dirty, and implacably ugly. Whatever was the attraction? The organizers of the 1978 exhibit queried alumni of the building and got illuminating answers. "Windows that open and shut at will of the owner!" (Martha Ditmeyer) "The ability to personalize your space and shape it to various purposes. If you don't like a wall, just stick your elbow through it." (Jonathan Allan) "If you want to bore a hole in the floor to get a little extra vertical space, you do it. You don't ask. It's the best experimental building ever built." (Albert Hill) "One never needs to worry about injuring the architectural or artistic value of the environment." (Morris Halle) "We feel our space is really ours. We designed it, we run it. The building is full of small microenvironments, each of which is different and each a creative space. Thus the building has a lot of personality. Also it's nice to be in a building that has such prestige." (Heather Lechtman)

In 1991 I asked Jerome Wiesner, retired president of MIT, why he thought that "temporary" Building 20 was still around after half a century. His first answer was practical: "At $300 a square foot, it would take $75 million to replace." His next answer was aesthetic: "It's a very matter-of-fact building. It puts on the personality of the people in it." His final answer was personal. When he was appointed president of the university, he quietly kept a hideaway office in Building 20 because that was where "nobody complained when you nailed something to a door."

Every university has similar stories. Temporary is permanent, and permanent is temporary. Grand, final-solution buildings obsolesce and have to be torn down because they were too overspecified to their original purpose to adapt easily to anything else. Temporary buildings are thrown up quickly and roughly to house temporary projects. Those projects move on soon enough, but they are immediately supplanted by other temporary projects—of which, it turns out, there is an endless supply. The projects flourish in the low-supervision environment, free of turf battles because the turf isn't worth fighting over. "We did some of our best work in the trailers, didn't we?" I once heard a Nobel-winning physicist remark. Low Road buildings keep being valuable precisely because they are disposable.

Building 20 raises a question about what are the real amenities. Smart people gave up good heating and cooling, carpeted hallways, big windows, nice views, state-of-the-art construction, and pleasant interior design for what? For sash windows, interesting neighbors, strong floors, and freedom.

Many have noticed that young artists flock to rundown industrial neighborhoods, and then a predictable sequence occurs. The artists go there for the low rents and plenty of room to mess around. They make the area exciting, and some begin to spruce it up. Eventually it becomes fashionable, with trendy restaurants, nightclubs, and galleries. Real-estate values rise to the point where young artists can't afford the higher rents, and the sequence begins again somewhere else. Economic activity follows Low Road activity.

Jane Jacobs explains why:

> Only operations that are well-established, high-turnover, standardized or highly subsidized can afford, commonly, to carry the costs of new construction. Chain stores, chain restaurants and banks go into new construction. But neighborhood bars, foreign restaurants and pawn shops go into older buildings. Supermarkets and shoe stores often go into new buildings; good bookstores and antique dealers seldom do. Well-subsidized opera and art museums often go into new buildings. But the unformalized feeders of the arts—studios, galleries, stores for musical instruments and art supplies, backrooms where the low earning power of a seat and table can absorb uneconomic discussions—these go into old buildings....
>
> Old ideas can sometimes use new buildings. New ideas must come from old buildings.[2]

A related economic sequence happened around houses. People used to store stuff in basements and attics (big tools and toys in the cellar, clothes and memories in the attic). These were the raw, undifferentiated, Low Roadish parts of the house. But after the 1920s, basements and attics were eschewed by new bungalows, Modernist homes, and ranch houses. Basement storage moved into the garage, but then it got displaced again when the garage

ca. 1940 - The electronics firm of Hewlett-Packard (revenues in 1992 of $16 billion) was founded in this garage in Palo Alto, California, in 1939 by William Hewlett and David Packard with a $538 loan from their electronics teacher at Stanford, Frederick Terman. The garage is now a state historical monument.

ca. 1973 - Three decades later, lightning struck another garage in Palo Alto, this one belonging to Steve Jobs. He and Steve Wozniak (a youthful Hewlett-Packard engineer) devised the first personal computer, the Apple, and another industry was born. In 1992 Apple contributed $8 billion to Silicon Valley's total sales of about $400 billion a year.

THE GARAGES OF SILICON VALLEY are no myth. And no accident. High-risk creative new directions in business are best taken by tiny start-up companies with no capital to spare for plant. They take root in buildings that no one else wants, like spare garages.

was converted to a studio, home office, spare bedroom, or rental unit. Where did the storage go next? Economic activity followed Low Road activity. The "self-storage" business took off in the

1970s and 1980s. Windowless clusters of garage-like spaces at the edge of town or edge of industrial districts were thrown together and rented out cheap.[3] In these spaces you find the damnedest things—a boxer working out, quiet adultery, an old gent in a huge chair enjoying a cigar away from his wife, an entire British barn in pieces, a hydroponic garden, stolen goods, a motorcycle repair shop, an artist's studio, someone shaping surfboards, lots of very ordinary storage, and, about once a month somewhere in America, a dead body.

Such trends are invisible to high-style architects, but commercial developers watch them closely. They noticed that small businesses often start up in garages, warehouses, and self-storage spaces, sometimes spawning whole Silicon Valley-type local boomtowns.

When my wife, Patty Phelan, started an equestrian mail order catalog business, she took over one bay of a huge old wood building left over from World War II—part of a shipyard that had built Liberty ships and tankers. Her bay had all the usual

2 Jane Jacobs, *The Death and Life of Great American Cities* (New York: Random House, 1961, 1993), p. 245. See Recommended Bibliography.

3 The Urban Land Institute reported: "With ever-increasing household mobility, a growing national preoccupation with possessions, and escalating demand for low-rent storage spaces (for records, data, and inventory) from businesses and professional offices operating out of relatively high-rent space, demand for self-storage is now equivalent to 2 to 3 square feet per person." "Self-Storage Adaptations," *Urban Land* (Oct. 1991), p. 28. In the early 1990s, self-storage facilities were beginning to include climate control, security, multistory buildings, and "acceptable" design on the exterior. It was a $2 billion-a-year industry, complete with its own trade magazine, *Inside Self-Storage*.

San Francisco Bay Model.
US Army Corps of Engineers.

1942 - The General Shops building was started July 20, 1942, and completed nine weeks later on October 3. That was typical of the pace at Marinship in Sausalito, California. Washington bureaucrats first raised the question of a San Francisco-area shipyard with Bechtel Corporation on March 2, 1942. Earth-moving and construction began in April, with the first building ready for occupancy on June 17. The first ship was launched that September. In three and a half years, Marinship built ninety-three Liberty ships and oil tankers. At peak production they were being completed one a day. The shipyard shut down at the end of the war.

21 December 1990. Brand.

December 1990 - My wife's mail order company, Phelan's, started in the bay behind the shed addition—184 Schoonmaker. In the course of five years she expanded the business into five spaces in the building. Very little of that activity was visible on the outside. There was a sign to draw people to the tiny shop in the shed, a flower box, a burgler alarm, and a picnic table for lunch in the sun.

21 August 1992. Brand.

August 1992 - The door acquired a little roof and became a Dutch door. The windows upstairs were refitted so they could be opened for desperately needed ventilation in the summer.

"INCUBATORS"—low-rent raw space—for start-up businesses were inspired by buildings like this. Left over from a World War II shipyard, the building is huge and rudimentary. Tenants rent narrow bays and are free to improve the space to suit their needs.

February 1987 - When a sculptor of neon moved out of the 184 bay in the old Marinship General Shops building, Patty Phelan and her one employee (pictured) moved in with their scheme for a catalog of "High Performance Horse Gear and Riding Apparel."

February 1992 - A carpenter friend of Patty's built stairs and a second floor in exchange for a fine saddle. She cut through the right hand wall into the adjoining bay and added a second floor in there, then cut through the back wall and expanded into three spaces sideways. In this original space, customer service is on the left, administration, design, and ordering upstairs, warehousing in the back, and shipping behind that.

Patricia Phelan.

24 February 1992. Brand.

amenities—concrete floor, a too-narrow, too-deep space, ill-lit, with a 60-foot ceiling. She and her staff froze in the winter and baked in the summer. But that space absorbed five years of drastic growth. The company went from one employee to twenty-four, from $50,000 a year to $3.2 million, while keeping all of its warehousing and shipping on the site. Piece by piece she grew the space, first constructing a second floor, then breaking through a wall into the adjoining bay when that tenant moved out and adding a second floor in there, then cutting through her back wall into some ceilingless interior rooms and roofing them in. Her rent stayed low while she added a skylight, ceiling fans, openable windows, a dutch door, lots more wiring, lots more lighting, and a kitchen.

That's the pattern that developers thought they might be able to duplicate—long, low, cheap building, a series of bays, each with a garage door, low rent, nothing fancy. Called "incubators," they were built by the hundreds, and they prospered. By 1990 there was a National Business Incubation Association boosting another Low Road-derivative industry.

The wonder is that Low Road building use has never been studied formally, either for academic or commercial interest or to tease out design principles that might be useful in other buildings. What do people do to buildings when they can do almost anything they want? I haven't researched the question either, but I've lived some of it. This book was assembled and written in two classic Low Road buildings. My writing office was a derelict landlocked fishing boat named the *Mary Heartline*. Decades ago, after its fishing career was over, a gay couple acquired it for dockside trysts, fixing it up like a Victorian cottage. Then two divorced gentlemen took it over, also for trysts, but it began sinking, so they

February 1993 - When Phelan's was sold and its operations moved to Illinois, the space was emptied, leaving behind improvements such as the greatly increased electrical wiring.

14 February 1993. Brand.

March 1993 - A month later a new mailorder start-up was in the bay, this one selling novelty products based on the "Doonesbury" comic strip. All the walls that Patty had cut through were healed up with new sheetrock, along with one new wall, and five assorted tenants were hewing the space to serve their assorted dreams.

8 March 1993. Brand.

May 1990. Brand.

FREEDOM IS CHEAP. Low rent equals high control if you're comfortable fixing up crude Low Road space. For me, discarded buildings became a luxuriously customized office (left) and library/studio (below).

July 1990. Brand.

1990 - Idyllic, rotten, the *Mary Heartline* had been used as an office for fifteen years before I got it in 1987. The raw wooden stairs had the kind of wear and patina one associates with ancient cathedrals. A secret garden had grown up around it, planted and tended by Kathleen O'Neill of the *Whole Earth Review* staff.

1990 - The self-storage container yard on Gate 5 Road in Sausalito, California, was the first of its kind in 1975 when Dick York was inspired by Moshe Safdie's Habitat building at the Montreal World's Fair to think about buying used Sea-Land containers ($500 apiece) and piling them up to make cheap, interesting space. People started renting them for self-storage, and he had a nice low-overhead business. In 1990 York (right) hired a crane to reconfigure the yard for more efficient parking. That's my library dangling up there.

January 1990 - Containers were invented in 1956 by Malcolm McLean of Winston-Salem, North Carolina, as a way to shortcut loading between ships, trucks, and flatcars. Made of aluminum, steel, and wood, they have a surprisingly comfortable space inside, and they close-pack efficiently with other containers.

March 1990 - I knew from editing *Whole Earth Catalogs* that the most important tool for organizing projects is lots of horizontal space and immediate-to-hand storage. Boat carpenter Peter Bailey built it cheap and sturdy. He told me I would regret using plywood for pinning up photos and other graphics on the walls, and he was right.

Brand.

Brand.

moved it onto land, ostensibly for repair. It became a real-estate office, a subscriptions-handling office, and then I got it. It was on no property map of the town. If you leaned against the hull in the wrong place, your hand would go through. It's probably gone by the time you're reading this.

Thanks to the gay couple's Victorian tastes, the place was a maze of little niches, drawers, and cupboards. It was like working inside an old-fashioned rolltop desk. One day I acquired a fax machine. There being no convenient place to park it, I used a saber saw to hack out a level place by the old steering wheel, along with a hole for the electrical and phone lines. It took maybe ten minutes and required no one else's opinion. When you can make adjustments to your space by just picking up a saber saw, you know you're in a Low Road building.

My research library was in a shipping container twenty yards away—one of thirty rented out for self-storage. I got the steel 8-by-8-by-40-foot space for $250 a month and spent all of $1,000

fixing it up with white paint, cheap carpet, lights, an old couch, and raw plywood work surfaces and shelves. It was heaven. To go in there was to enter the book-in-progress—all the notes, tapes, 5 x 8 cards, photos, negatives, magazines, articles, 450 books, and other research oddments laid out by chapters or filed carefully. When the summer sun made it too hot for work, I sawed a vent in the wood floor, put a black-painted length of stovepipe out of the ceiling, and slathered the whole top of the container with brightly reflective aluminum paint—end of heat problem. That's how Low Road buildings are made livable: just do it.

In fact, weather becomes a perverse attraction. Whereas competent sealed buildings lull us with their "perfect" climate, and incompetent ones drive us crazy with their uncontrollable heats and colds, a drafty old building reminds us what the weather is up to outside and invites us to do something about it—put on a sweater; open a window. Rain is loud on the roof. You smell and feel the seasons. Weather comes *in* the building a bit. That sort of invasion we would condemn in a new building and blame the architect, but in a ratty old building—designed for some other use, after all—there's no one to blame.

Such buildings leave fond memories of improvisation and sensuous delight. When I lived with an artists' commune in an old church in New York state, I slept in the steeple in front of the rose window overlooking the stream below. The major problem was being pooped on by pigeons, so I made a canopy from the canvas of a large bad painting (art side up) and thereafter slept in comfort, cooed to my rest by flights of angels.

Low Road buildings are peculiarly empowering.

February 1993 - People asked, "How can you stand it in there without windows?" All I could say was, "A library doesn't need windows. A library is a window." In February I was using the flat space to organize Chapter 12 with the 5 x 8 cards on which all of the book's raw research was taped. By this time I had followed Peter Bailey's advice to have sheet steel on the walls, and little magnets holding up the photos.

Brand.

CHAPTER 4

Houseproud: The High Road

THE DUCHESS of Devonshire—born Deborah Mitford, the youngest of the celebrated Mitford sisters of Britain (Nancy, Jessica, Diana, Unity)—has the family talent for writing and for getting into interesting trouble. In 1941 she married Andrew Cavendish, the 11th Duke of Devonshire, and in 1958 the couple took on the reviving of Chatsworth, a spectacular country house regarded as one of the "treasure houses" of England, which had been in the family since it was built in the years 1686-1707. In 1958 such a task was considered impossible and thankless.

No wonder. The main building—the "house"—had 175 rooms, 51 of them enormous, and it needed serious work. Set in the beautiful Peak District of Derbyshire, it was surrounded by miles of park landscaped in the 18th century by "Capability" Brown, and also in need of endless maintenance. The saga of how the family and estate staff succeeded in making Chatsworth work for the benefit of all—320,000 visitors a year—is interesting in its own right, but more to the purpose of this chapter is understanding why it was worth the trouble.

Deborah Devonshire wrote in 1982:

> The house looks permanent; as permanent as if it had been there, not for a few hundred years, but for ever. It fits its landscape exactly. The river is the right distance away and the right width. The bridge is at a comfortable angle for leaning and gazing from. The stone from which the house is built comes out of the ground nearby, and so it is the proper color, on the bird's nest theory of building materials being at hand and of the place, and thereby right for the surroundings....

Christopher Simon Sykes. From a gorgeous book, *Ancient English Houses* (London: Chatto & Windus, 1988), p. 135. See Recommended Bibliography.

ca. 1985 - BEELIEGH ABBEY, near Maldon, Essex, has been a private home since 1536. This doorway tells its own history. In a 13th-century wall, arched windows were cut in the 15th century. This window was bricked in—apparently in two stages—and oak-framed in the 17th century for a doorway. The battened door itself is late 19th-century. High Road buildings, reshaped and refined by centuries, flaunt their checkered histories.

It has taken more than four hundred years to reach this apparently effortless state of perfection by a combination of good luck and good management. The luck is that every generation of Cavendishes has loved and respected the house and its surroundings, and each has added something to it, and the good management is literally so in that no house has been better served by its stewards and housekeepers than has Chatsworth....

The charm, attraction, character, call it what you will, of the house is that it has grown over the years in a haphazard sort of way…. Each room is a jumble of old and new, English and foreign, thrown together by generations of acquisitive inhabitants and standing up to change by the variety of its proportions and the strength of its cheerful atmosphere. Likewise the outside. There is something surprising to see wherever you look; nothing can be taken for granted….

It has all been done by the confident dictators of yesteryear, with no recourse to committees or wondering what other people will think about their additions and subtractions.[1]

Those are the basics of what makes a High Road building acquire its character—high intent, duration of purpose, duration of care, *time*, and a steady supply of confident dictators. In time such a building comes to express a confidence of its own. I interrupted the Duchess with some questions one day on her daily round of tending to the rooms, the garden, the staff (90 in summer), the shops, the tea room. She said, "This house has such a personality it imposes itself on you in a funny way. There's a kind of discipline which you feel when you've been surrounded by these rooms as long as I have." I asked what kept her going. "The love of it. That's all. You've got to be in love with it to make it work. It's like having a job. You just go on because you're fascinated by it."

Besides gaining the loyalty of their occupants and visitors, old buildings that stay in use rise to other freedoms. By spanning generations, they transcend style and turn it into history. The tiresome furniture of one's grandparents is put into storage to be rediscovered by one's grandchildren. The Baroque wing that once was thought marvelous, then became risible, now is honored for its contrast to later additions. By showing a tangible deep history, the building proposes an equally deep future and summons the taking of long-term responsibility from its occupants.

Chris Barker. Cover photo for *The House*. See footnote below.

ca. 1980 - CHATSWORTH and its currently resident Duke and Duchess of Devonshire. The west facade of the house was designed by architect Thomas Archer and the First Duke, William Cavendish, around 1705. It has been open to the public since then. Deborah Devonshire describes the responsibility: "If you live in the same place for a long time you become hefted to your hill like an old sheep…. If your surroundings happen to be renowned for their beauty, bringing people from all over the world to see for an hour or two what you see every day, and if you sometimes have the opportunity either in some way to add to that beauty or to prevent something dire happening, then you are indeed lucky."

Such agglomerations become highly evolved, refinement added to refinement ("The bridge is at a comfortable angle"), the sensible parts kept, the humorous parts kept, the clever idea that didn't work thrown away, the overambitious conservatory torn down, the loved view carefully maintained, until the aggregate is all finesse and eccentricity. The measure of successful evolution is intricate vivacity.

But there are substantial disadvantages as well to life in a High Road building. Occupants in a lean time can be crushed by trying to maintain what was built in fat times. (In the lavish Victorian era, English country houses expanded to insupportable scale.)

---

1 Deborah Devonshire, *The House* (London: Macmillan, 1982), pp. 15-16, 226. See Recommended Bibliography.

36

1754 - Salisbury's nave is 80 feet high. The original choir screen (or pulpitum) in the center foreground was at first brightly painted and had sculptures of kings in its niches with angels above them, but political events of the mid-16th-century (the Dissolution and the Reformation) removed them. The organ dates from 1661.

ca. 1865 - In 1787 the famous architect James Wyatt (later infamous for overzealous remodeling) was hired by a wealthy bishop to redo Salisbury. The orignal screen and old organ were replaced with a new screen by Wyatt and a new organ by Samuel Green. Bright medieval painted stone walls and vaults were painted over in various shades of "stone."

1971 - The exterior of Salisbury Cathedral looks now much as it did in 1266, except for the tower and spire, which were added in the 14th century. Weighing 6,300 tons, they threatened the whole structure, but reinforcements and additional buttressing contained the damage to minor distortion. At 394 feet, it is the tallest medieval stone spire in the world.

15 October 1929. Royal Commission on the Historical Monuments of England. Neg. no. BB84/1412.

1929 - As late Gothic Revival supplanted early Gothic Revival, the next restorer undid the work of the last one. In 1858 George Gilbert Scott (later Sir) was hired to rethink Salisbury. The organ was moved, Wyatt's screen was demolished and replaced by an open metalwork one (by Francis Skidmore) in 1876, and a Scott-designed pulpit appeared on the left in front of it, along with an ornate Scott reredos behind the altar in the background.

16 November 1965. Royal Commission on the Historical Monuments of England. Neg. no. BB65/4372. The Salisbury interior sequence is from Gerald Cobb's *English Cathedrals: The Forgotten Centuries* (London: Thames and Hudson, 1980), pp. 120-121.

1965 - In 1960 a new theory of English cathedrals declared that they should present an uninterrupted vista from end to end, and Scott's screen and reredos were swept out of Salisbury. Only his pulpit remained. For the time being.

CONSTANT REVISION is the fate of even the most institutional and High Road of all buildings— religious edifices such as Salisbury Cathedral (1266).

Whatever grows to splendor becomes a target for levelers with punitive taxes and casual confiscations. And updating the services of any building that once had "perfect" amenities, because it could afford to, is a major trial. Try putting modern plumbing and heating into a stone Chatsworth—it's like performing lung surgery on a tetchy giant. The High Road is high-visibility, often high-style, nearly always high-cost.

Whereas Low Road buildings are successively gutted and begun anew, High Road buildings are successively refined. These are precisely the two principal strategies of biological populations—the opportunist versus the preserver: "r-strategy" versus "K-strategy" in the jargon.[2] It is the difference between annual and perennial plants—between weeds like dandelions which scatter profuse seed to the winds, and dominant species like oak trees, which nurture their few acorns and build an environment that protects the next generation. Individuals of opportunistic species are typically small, short-lived, and independent, putting all their energy into productivity. Preserver species are more often large, long-lived, densely interdependent and competitive, rationing their energy for high efficiency.

The sustained complexity of High Road buildings leads in the fullness of time to rich specialization. They cannot help becoming unique. They respond to so many hidden forces, they are in part mysterious, sustained by subtleties. At the same time they are filled with obsolete oddities, preserved out of habit until odd new uses are found for them. (Where can we string the new fiberoptic cable? How about in the old laundry chute?) High Road buildings are common, but the points I want to make about them are best demonstrated in extreme examples.

The divergent uniqueness of High Road buildings can be observed perfectly in three parallel buildings from American history, the country homes of George Washington, Thomas Jefferson, and James Madison—America's first, third, and fourth Presidents. Each is a classic "biography" house, having closely sustained and reflected the changing life situation of the main

occupant and his family. All in the state of Virginia within a few hours drive of each other, this trilogy of biography houses is open to the public. Each is a house on a hill, a "mont"—Washington's Mount Vernon, Madison's Montpelier, Jefferson's Monticello. Each served an identical function as the headquarters and manor house of a working plantation. Each is well preserved because the soil was exhausted by the early 1800s, the plantations failed, and little further alteration could be afforded.

Their owners were so close in time, place, career, and station—they were colleagues and friends—that you would expect their homes to be much alike. Indeed, with their classical pedimented symmetry, the buildings look glancingly alike now, but they *grew differently*. Facing identical situations, the homes were shaped by distinctly diverse characters and lives.

George Washington's place, when he began expanding it in 1758, was a one-and-a-half-story cottage built by his father twenty years before. The reshaping would continue to his design for thirty years. To make room for his new wife Martha and her two children (and employing her money), he nearly doubled the space by raising the house a story. After adding plantation outbuildings—called "dependencies" in the South—in 1775 he devised a grand scheme for Mount Vernon.

The plan went forward during the Revolutionary War years (1775-1783), while he was in command of the Continental forces, losing

---

2 In population-biology equations "r" is rate of population growth and "K" is carrying capacity. In a new or disturbed environment, opportunist "r-selected" species are at an advantage because they can grow rapidly to occupy the new niches available. Preserver "K-selected" species do best in a stable ecosystem, where their efficiency and tenacity keep them secure in the niches they have filled, often in complex interactivity with other species. Consequently, Low Road buildings can prosper in the chronically disturbed economic environment of cities, while High Road buildings thrive best in the more stable countryside.

3 William Seale is author of the definitive book on the White House, *The President's House*. See Recommended Bibliography.

4 Conover Hunt has helped The National Trust for Historic Preservation with the preservation and historic understanding of Montpelier. Her book is *Dolley and the Great Little Madison* (Washington, DC: AIA Foundation, 1977).

27 November 1991. Brand.

1991 - MOUNT VERNON'S PIAZZA (double-story porch) was an original idea of Washington's. Also new in the region was his siding of pine planks beveled and painted with sand to look like stone masonry— they rotted more quickly, unfortunately. On the left is one of the "dependencies," the kitchen, linked to the main house by a curving palisade. A better illustration of the charm of the piazza would have been a 1797 sketch by Benjamin Latrobe of Washington and friends having tea and admiring the view of the Potomac, but the sketch is privately owned and is controlled by the curator of Mount Vernon. Though the illustration has appeared in other books, she decided that this book was inappropriate for it.

battles to the British but winning the war and the new nation's adulation. He expanded the building at both ends, adding a personal library/office and an upstairs bedroom at one end and a grand banquet room—two stories high, to handle ceremonial demands—at the other. In hospitable Virginia, country manors were treated like inns by travelers. Washington made the one end of his inn carefully private, with a back stairs from bedroom to office, while the other end welcomed visitors.

It would have been a long skinny house but for a brilliant innovation of his, perfectly suited to the building and the site. Mount Vernon overlooks a beautiful stretch of the Potomac River. For the side of the house facing the view, where there is morning sun and afternoon shade, Washington invented a two-story-high porch ("piazza") extending the whole length of the house and unifying it. Flagstones made it impervious to rain and rot. To this day it is one of the nicest places in America to just sit. Would Washington have known how to make the piazza so amenable if he hadn't lived in the house for fourteen years first?

"I think Mount Vernon is the best added-on-to American house of the 18th century," comments William Seale, a building historian closely familiar with the place.[3] "Washington made it cohesive. He did it as he needed it, and he was a man totally without pretense. I think Mount Vernon is so wonderful because you can follow the man's logic. He is always a surprise. You think he's a lumbering sort of oaf, and he's not. He turned out to be very wise, often much wiser than Jefferson, much more feet-on-the-ground. When Washington went on to build the White House, he had a stately home in mind there too. He put the East Room where he put that banquet room at Mount Vernon, at one end. The White House is zoned very much like Mount Vernon."

A different design intelligence was manifest in James Madison, regarded as the leading genius behind the US Constitution. Like Washington, in 1797 he moved a brand-new wife and stepchild into a house built by his father. Unlike Washington, his parents were still living there. So began "a wonderful duplex story," in the words of Conover Hunt, biographer of James and Dolley Madison.[4] The design problem was this: the elder Madisons had a traditional central-hall "double-pile" two-story house, which they aimed to stay in; but James was due to inherit the plantation and metal foundry—he was beginning to take over management—and he needed room for his new family.

Solution: expand the house on one side by thirty feet, move the central entrance and hall over a notch, and gussy up the new symmetry with a pedimented portico. In effect, a second home was added under the same roof, with clear division between the two households. Each had a separate kitchen and separate servants; they kept separate schedules; and the two sides of the house were physically independent, with only a single connecting doorway upstairs. This bicameral arrangement kept peace in the family and gave some flex—either side could borrow use of rooms from the other when needed. (A similar division keeps Chatsworth livable: the family lives privately on the middle floor; the public, coming in through another entrance, tours daily through the ground and top floors, unaware that they're missing anything.)

**George Washington's
MOUNT VERNON**

**James Madison's
MONTPELIER**

1758

1765

1774

1798

1799

1812

1935

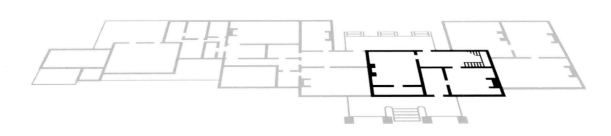

THESE DRAWINGS by Donald Ryan, all to the same scale, show the changes to their homes made by Washington, Madison, and Jefferson (plus the Du Pont alterations of Madison's Montpelier). Existing extensive documentation of each building made the detailed comparison possible.

The heavy lines in the floor plans indicate walls of the original buildings.

SUPERFICIALLY SIMILAR, DEEPLY DIFFERENT, the homes of Washington, Madison, and Jefferson expressed differing intelligences responding to differing circumstances.

Thomas Jefferson's MONTICELLO

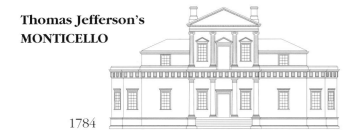

1784

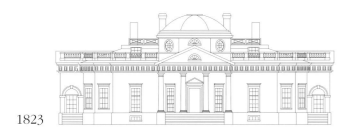

1823

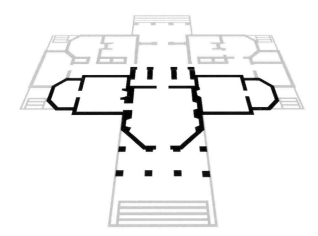

The first drawing of Monticello shows the east facade, which was demolished by Jefferson after 1793 and became the middle of the house when he doubled its size to the east. The drawing shows what Jefferson planned, but it probably was not completed when he tore it down. The second drawing is of the west facade, the side with the dome, which continues to look the same today (as in the photo, right).

1991 - MOUNT VERNON has subtle asymmetries in its facade that reflect its genesis. The windows offset to the right of the front door show where Washington's father placed the central hall stairs. Since the banquet hall at the left of the house is two stories high inside, the upper two windows on the left are fake—placed there just to preserve visual symmetry.

ca. 1884 - MONTPELIER as it looked after 1812, with wings added to handle the entertainment duties expected to follow Madison's final term as president. The wing on his mother's side of the house (right) was built mainly for symmetry.

ca 1937 - MONTICELLO's west facade developed by stages of ornament rather than size. This is the original side of the house (1784), with Jefferson's private quarters on the right. The dome was added in 1801 after his return from France. The portico was not fully completed till 1823.

10 May 1991. Brand.

1991 - MONTPELIER, after the du Ponts were done with it, was a mansion of fifty-five rooms. Now a museum house run by the National Trust for Historic Preservation, two du Pont rooms have been kept furnished in the grand manner of the 1910s and 1930s. The rest of the building is being restored to exhibit its three stages of development by the Madison family.

In 1809 Madison faced a new design problem with the house. He had just been inaugurated as president, and he knew from what had happened to George Washington that he was going to be inundated with visitors once he retired. Virginia hospitality required entertaining them, and Dolley was a famous hostess. Meanwhile his mother showed signs of living forever (she lived to 97), and he wanted her left undisturbed in her side of the house. And he depended on Montpelier as a retreat from his presidential labors.

This time Madison hired professionals, two builders fresh from several years working on Jefferson's Monticello. Their job was to add single-story wings at each end of the house, redo the portico, add a colonnade on the other side, and reconfigure many of the interior rooms. In Madison's library/study over the entryway they were to improve the window to match the view. Conover Hunt: "That was the intellectual heart of the house. It has a fabulous view. It's a very contemplative room, it's so plain." The expanded house served Madison's retirement after 1817 perfectly. There he continued as a much-consulted sage helping interpret the Constitution until his death in 1836.

A series of later owners made minor modifications to Montpelier

until 1901, when a branch of the du Pont family backed up a truckload of money and made major changes, nearly tripling the house in size. By the 1930s it had become a center for fashionable equestrian events on the grounds (the low-ceilinged space in the top of the portico became "the Jockey Room.") Conover Hunt: "The whole house is a classic example of jackleg architecture. If you want to see a place that expresses its owners, that's the building. It's a house within a house within a house within a house. All its owners are expressed—beam by beam, wall by wall."[5]

James Madison's best friend and lifelong co-conspirator was Thomas Jefferson; they even used a private code in their correspondence. Monticello was only 28 miles away from Montpelier, close enough for Madison to buy Jefferson's nails and politely ignore his architectural advice.

Monticello was one man's fantasy right from the start, and it remained a barely inhabitable fantasy almost to the end. The man who wrote the Declaration of Independence was determined not to copy standard English country houses the way the fathers of Washington and Madison had. A creature of the Enlightenment, Jefferson looked to classical scholarship, especially the neo-Roman designs of Andrea Palladio and Robert Morris. All his life he was an energetic evangelist of neoclassical taste to the rough new republic. His home was his primary laboratory and model, and it remains his best monument—"the most creative artifact of an exceptionally creative man," writes Jack McLaughlin, the author of *Jefferson and Monticello: The Biography of a Builder.*[6]

---

[5] The du Pont family eventually turned Montpelier over to the National Trust for Historic Preservation, which opened the building to the public in 1987. In 1992 Mount Vernon was drawing a million visitors a year, Monticello half a million, and Montpelier 40,000—expected to increase rapidly.

[6] Jack McLaughlin, *Jefferson and Monticello* (New York: Holt, 1988), p. vii. See Recommended Bibliography.

[7] Above title, p. 14.

ca. 1890 - MONTICELLO. Jefferson's reverence for Palladian symmetry was overcome by his craving for privacy when he added louvered "porticles," as he called them, to the private end of the house (left). They screened his study and bedroom from prying eyes (strangers would stand around outside watching him), and one served as an aviary. A later owner removed them; they have not been restored. This east facade of the house was designed by him to look like one story, while it was in fact two, as shown by the two-leveled window next to the east portico (right).

1912 - "No building code in America would permit such staircases to be constructed," writes Jack McLaughlin. "They are built into stairwells a scant 6 feet square, so small that the stair treads are only 24 inches wide and the risers dangerously high. Because the staircases turn twice on each floor, the stairway is virtually spiral, with hazardous, narrow, wedge-shaped steps." (*Jefferson and Monticello*, p. 5.)

McLaughlin, himself an owner-builder, sums up what happened at Monticello:

> Most owner-builders take inordinate lengths of time to complete their projects; Jefferson took fifty-four years. Many owner-builders construct dwellings larger than necessary; Jefferson, a widower, built a thirty-five-room mansion. Owner-builders invariably extemporize as they build, adding to and modifying their original design as the house grows. Jefferson built one house, tore much of it down, doubled its size, and continued to alter, remodel, improve, and add to it for decades. It is a wonder that the house was ever finally completed; many thought it never would be.[7]

Jefferson was twenty-five in 1768 when he began preparing a remote and beautiful mountaintop for his dream house—the first of many decisions for the aesthetic over the practical. Untraveled, he relied on the pattern books of Palladio and Morris for layout and details as he began making meticulous drawings of his ideas. Having been burnt out of a wood family home, he wanted to build in brick and stone for the ages. Only a small brick outbuilding had been completed by 1772 when Jefferson moved in with his new bride, Martha. During the ten years of their marriage, their home was a chaotic construction site as well as the headquarters of a busy plantation. One year their daughter Maria fell into the basement. Years later on her honeymoon at Monticello, she sprained her ankle tumbling out a doorway. While working late in the unclosed house one night, Jefferson had to stop writing because his ink had frozen. After his wife's deeply mourned death in 1782, he never married again, and Monticello became a widower's hobby. As Conover Hunt puts it, there was no wife to declare, "You pull down one more wall and I'm out of here."

Monticello's design evolved steadily as Jefferson gained in experience. It was after he had completed the foundation that Jefferson decided to expand the two ends of the building with half-octagons—a distinctive but awkward form that fascinated him for the rest of his life. In 1789, he came back from five years as ambassador to France with a headful of the latest architectural ideas. He wanted a dome. A famous man now, he needed more room to entertain and private space to hide from guests. Monticello's ingenious, idiosyncratic design can be read as an elaborate effort to do two things: to protect Jefferson's privacy—by bringing in light without visibility to prying eyes—and to honor Palladian formalism. Monticello expanded sideways, in mirror image of the original, with rooms separated by narrow hallways and even narrower stairs (too steep and dangerous for tours nowadays). The roof became hideously complicated, and leaked. Jefferson experimented with wood shingles, lead, copper, and tin-plated wrought-iron roofing.

By the time Jefferson retired in 1809 after two terms as President, his home had an idyllic bedroom-library-greenhouse suite on the south wing, walled with his 6,000 books, but the construction saga was still unfinished. Final completion of the impressive porticos did not occur until 1823, three years before his death. Jack McLaughlin points out that the splendidly preserved Monticello that visitors now see never occurred in Jefferson's lifetime. It was fresh but unfinished in 1809, finished but badly decayed in 1823—Jefferson had no money to pay for maintenance (he sold his books to form the core of the Library of Congress collection).

Reading these three fiercely distinct building autobiographies, what do we learn about the authors? We find in Washington a conservative revolutionary, with sensible ideas (the banquet room) and delight in a bold stroke (the two-story piazza); an accommodating Madison (wife here, mother there) improvising on vernacular wisdom; a fussily artistic Jefferson (busy octagons, perilous stairs) building in high idealism for a distant polity.

This is the way to grow a High Road building. Take it by stages, with constant minute refinement and breezy innovation comfortably expressed by the attentive intelligences coevolving with the building. The result is human: a building by the people, for the people, and of the people within.

For contrast, watch what happens when institutions try to make High Road buildings, which they usually do. Institutions aspire to be eternal, and they let that ambition lead them to the wrong physical strategy. Instead of opting for long-term flexibility, they go for monumentality, seeking to embody their power in physical grandeur. Post offices, colleges, and state capitols belie and hinder their high-flux information function with stone walls, useless columns, and wasteful domes. The building tries to stand for the function instead of serving it. The White House *is* the executive branch; the Pentagon *is* the US military.

Actually the Pentagon is an interesting counter-example. It is a highly functional and relatively unpretentious office building. I have great admiration for military buildings. Economical and practical, they are innocent of architectural guile, well maintained, and highly adaptive in a Low Road way. Unlike most institutional buildings, they have a clear and immediate responsibility structure driving their use, adaptation, and maintenance. The career of the post engineer is over if he leaves the buildings in his charge in worse shape than he found them. Military buildings are rich, not in money, but in people with time to work on them. The same is true of fire houses: between alarms, firefighters are always busy around the place.

But most institutions occupy blocked High Road buildings. A frozen bureaucracy and a frozen building reinforce each other's resistance to change. Change is obligatory, since bureaucracies always grow, but responsibility is dispersed and delayed in a maze of anxious responsibility-avoidance, and the building sits heavy and inert. Near any institutional building more than a decade old, you are likely to find a host of clumsy Low Road expediencies—trailers, temporary add-ons, people working in windowless storage rooms, space rented in nearby commercial buildings.

Libraries are a glorious case for study. They exude architectural permanence. Meanwhile their collections grow and grow, and the pressure builds. Do any expand gracefully? The Library of Congress does not. The modern world's greatest library, by 1990 it had accumulated 97 million items—including 15 million books (7,000 of them just on Shakespeare), four million maps, 14 million photographs, 40 million manuscripts, and a vast international collection including the largest body of Russian material outside Russia. Seven thousand items flood in to the collection every working day—1.5 million a year. They are housed in three huge buildings on Washington's Capitol Hill, plus many remote warehouses.

"I think of this place as a chambered nautilus," says Craig D'Ooge, information officer for the Library. The first chamber, the gaudy

Jefferson Building, was the largest library structure in the world when it was finished in 1897. Critics said it would never be filled. They should have said the opposite. It was badly overflowing when a second chamber, the plain Adams Building, was added in 1939, doubling the available space. A third doubling had to be added with the starkly Modernist Madison Building in 1980. D'Ooge: "There's a huge lag time between when you need something and when you actually get it. We were bursting at the seams when we asked for the Madison Building in *1958*." That twenty-two-year lag time between stimulus and response, as filtered through Congress, indicates a barely functional feedback loop of governance of one of the nation's most invaluable information-age assets. Because it is not allowed to anticipate its growth realistically, this superb institution barely functions. Most of the collection is neither processed nor accessible.

16 August 1991. Brand.

1991 - THE LONDON LIBRARY at 14 St. James's Square. The modest facade and homey Reading Room (right) date from 1898—London's first steel frame building. One hero of the library, according to the members' guidebook, was Frederick Cox, who served at the issue desk for forty years: "He was reported to know every book in the library and its exact place on the shelves. He certainly knew every member by name and interests, and often could tell them, better than they knew themselves, what book it was they wanted." The guidebook advises members, concerning the use of the Reading Room, "If you go to sleep in the armchairs you will not be disturbed till closing time."

Library of Congress. Neg. no. PG-5475 C.

1969 - THE LIBRARY OF CONGRESS had so outgrown its original two buildings by 1969 that staff had to crowd their offices around the statues on the decorative balconies overlooking the Main Reading Room in the Jefferson Building (1897). This photo helped convince Congress that a third building really was needed.

For an unfair but revealing contrast, look at the fairly graceful evolutions of Boston's Athenaeum and Britain's London Library. As private libraries, both are directly responsive to their users. Though institutional, they are treated exactly like High Road homes, with an affection and attention to detail that grows through the centuries. Both are filled with the quirks and traditions characteristic of long, independent life.

Their histories are parallel. Founded in 1807, the Boston Athenaeum grew through three different locations, then settled in 1849 in its present building, built for the purpose on fashionable Beacon Street. The London Library was founded by historian and essayist Thomas Carlyle in 1841 and moved to its permanent site, a townhouse on St. James's Square, in 1845. The Boston Athenaeum kept ahead of its growing collection by a series of convolutions and convulsions within its space—replacing a fancy staircase with stacks for 90,000 books, adding internal floors,

ca. 1853 and **1840** - Founded in 1806, the Athenaeum's second and fourth buildings in Boston were Scollay's building (top, foreground), where it resided from 1807 to 1809, and the James Perkins house (bottom), occupied from 1822 to 1849.

1849 - In 1847 a competition was held to decide the design of the Athenaeum's fifth (and so far ultimate) building. Perhaps because it was a blind competition (designers' names not revealed), it was won by a non-architect, Edward Clark Cabot. His neo-Palladian palazzo facade became the most permanent feature of the library.

1852 - Behind its apparently two-story facade at 10 1/2 Beacon Street, the building had three stories inside—an art gallery on top, a sculpture gallery on the ground floor, and the library and reading room in the middle. Watch what happens with the surrounding buildings.

THE BOSTON ATHENAEUM's internal convulsions to keep up with its growing book collection stayed hidden behind a facade of High Road decorum. It exuded institutional foreverness as the city flickered around it.

1915 - The additional two floors were set back so they would not be visible from the street or interfere with Cabot's sacred facade. The ghost carriage evidently departed while the exposure was being made.

1913 - Much of Boston burned in 1872. Fear of fire led to a major renovation and partial fireproofing of the Athenaeum in 1905. Warned that it was still a firetrap, the trustees ordered the entire building gutted in 1913, and it was reframed with steel (visible through large window). A sub-basement was added, along with two additional floors on top, and the interior was redesigned—except for the reading rooms, which were rebuilt exactly as before.

1895 - Already two renovations had reorganized the interior—the books invaded the top and bottom floors, and an extravagant staircase was removed to make room.

1902 - With tall new buildings on each side, the north (street) side of the Athenaeum was declared too dark inside for reading, so the stacks were moved to that side, and reading rooms were developed on the south side, looking over the historic Granary burying ground.

1990 - The spots of light on the Athenaeum are sunlight reflected from the windows of surrounding skyscrapers. The last time I talked to the director of the Athenaeum, he said the trustees were trying to figure out where to expand the library next—perhaps out into the setback space in front of the top two floors.

1990 - Though from the street the Athenaeum still appears to be two stories high, a look down the stairwell reveals ten levels—including two basements and the mezzanine levels for stacks in the high-ceilinged main floors.

1976 - A quiet facelift for the Brahmin of Beacon Street brightened the dark stone, but not too much, and repaired window and roof problems.

48

1915 - Following the major rebuilding in 1913-1914, the new top floor reading room was designed by architect Henry Forbes Bigelow to harmonize with the earlier reading rooms on lower floors. It surpassed them.

1940 - Though intensely used, the scene remained the same. Books moved around on the shelves.

1990 - After 75 years, same as it ever was. It's where I go to write when I'm in Boston.

ATHENAEUM INTERIOR. Like most conservative High Road buildings, the Boston Athenaeum has some rooms that remain reassuringly the same for decades. And some—less public—that change constantly to match the changing work they have to do.

One reason the Fifth Floor Reading Room is so agreeable is that it was arranged on the alcove system popular in the 19th century. Later Carnegie libraries got rid of alcoves so the librarian could keep an eye on everybody.

Sally Pierce, the Athenaeum's Prints Librarian, says that the reading rooms are like "stage sets"—carefully maintained to stay the same. They are frequently photographed, but the backstage rooms almost never are, and they're where all the action is—the offices jammed in hidden corners, the kitchen, the laundry room (for the Wednesday Tea linens), and the basements.

1927 - The new sub-basement excavated in the 1913-1914 gut-and-rebuild.

1990 - The same area of the sub-basement: lots more shelving, more electrical wiring, the door has been narrowed from double to single, a new concrete floor has been laid in Room 3, and compact shelving was installed in there.

putting in compact shelving (a recent invention that lets shelves roll together in a solid mass). The London Library clung to its site by growing discreetly into adjoining buildings. Each library was an early experimenter with steel-frame bracing to handle the massive volume of books—700,000 in the Boston Athenaeum, one million in the London Library, both collections growing by 7,000 volumes a year. In the 1990s, both were continuing to expand quietly into their nearby airspace.

The product of careful continuity is love. Members of both libraries adore their buildings. A poet trustee of the Boston Athenaeum declared, "No other Boston institution has anything like its unique, endearing and enduring atmosphere. It combines the best elements of the Bodleian, Monticello, the frigate *Constitution*, a greenhouse, and an old New England sitting-room."[8] A journalist member of the London Library describes the place as "a pleasing oasis, all leather-bound and beeswax-scented, a hallowed temple of peace and civility, of scholarship and serendipity."[9] In both libraries the stacks are open, nearly any book can be borrowed and taken home, and the London Library has an extremely busy Suggestions Book right by the door. The directors answer to nobody but their customers, and they answer quickly. Trust, intimacy, intense use, and time are what made these buildings work so well.

They are not distinguished-*looking*. What such buildings have instead is an offhand, haphazard-seeming mastery, and layers upon layers of soul. They embody all the meanings of the word "mature"—experienced, complex, subtle, wise, savvy, idiosyncratic, partly hidden, resilient, and set in their ways. Time has taught them, and they teach us.

---

8 David McCord, quoted in *Change and Continuity: A Pictorial History of the Boston Athenaeum* (Boston Athenaeum, 1985), p. 5.

9 Simon Winchester, "Every Member's Private Library," *Great Escapes* (Oct. 25, 1987), pp. 22-25.

10 Robinson Jeffers, "Tor House," *The Collected Poetry of Robinson Jeffers*, 3 vols. (Stanford, California: Stanford Univ. Press, 1988), vol. I, p. 408.

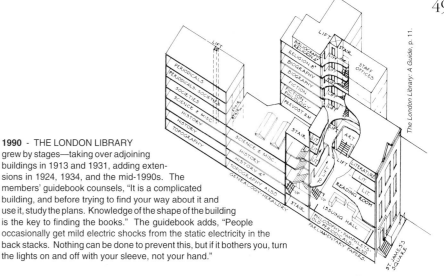

1990 - THE LONDON LIBRARY grew by stages—taking over adjoining buildings in 1913 and 1931, adding extensions in 1924, 1934, and the mid-1990s. The members' guidebook counsels, "It is a complicated building, and before trying to find your way about it and use it, study the plans. Knowledge of the shape of the building is the key to finding the books." The guidebook adds, "People occasionally get mild electric shocks from the static electricity in the back stacks. Nothing can be done to prevent this, but if it bothers you, turn the lights on and off with your sleeve, not your hand."

The London Library: A Guide, p. 11.

Many High Road buildings are very high-cost, but since maturity can't be bought, some of the best manage almost without money, lavishing time and attention instead. The California poet Robinson Jeffers and his wife Una were impoverished all of their lives, though he established a world reputation with his defiantly unmodern poems and plays. Together they built a distinctly unmodern house in Carmel, on the Pacific coast south of San Francisco. Made of granite boulders from its beach, Tor House is a poem-like masterpiece. It may express more direct intelligence per square inch than any other house in America.

The couple bought the small piece of land for $200 in 1918. Una was a lifelong student of medieval Irish towers. Robin apprenticed as a mason with the contractor who built the first part of their house until "my fingers had the art to make stone love stone;"[10] then he did all the rest of the construction himself while living in the house and raising twin boys. He studied forestry and planted trees. As a gift for Una, he spent four years building a massive tower of stone. Hawk Tower has walls four feet thick and is forty feet high, with two stairways—one outside, one thrillingly narrow and secret inside—two fireplaces, an oriel for admiring the sea's weather, and five stories if you include the dungeon for the twins to play in.

49

© Morley Baer. Neg. no. 82192-2.

1964 - TOR HOUSE (center) was built by Carmel contractor M. J. Renolds to Una Jeffers's design, with poet Robinson working as a day laborer. They hauled the granite stones from the Pacific beach a hundred yards beyond the house. Of Hawk Tower (left) Jeffers's son Donnan wrote in 1971, "Some of the stones he used were of enormous size and weight, and looking at the structure now, it seems incredible to me that one man could have done the job entirely unaided, especially in the relatively short time of four years." To the right of the house is the dining room that Robin added in 1930.

29 July 1991 : Brand.

1991 - The middle floor of Hawk Tower was designed as Una's room—a mahogany-paneled, medieval-feeling place for reading, writing, or playing the melodeon. At one end of the room is this three-windowed oriel overlooking the ocean. An admirable place to watch the arrival of Pacific storms.

Mortared prominently into the walls of the tower and the house are mineral relics from the entire sacred and natural world—bits from all of the towers in Ireland, including Yeats's Balylee, and from castles and cathedrals of England and Europe, the Babylonian temple of Uruk, the Great Wall of China, and the Great Pyramid of Cheops; a piece of Hadrian's villa and a piece of California's Lick Observatory; lava from Mount Vesuvius and from Mount Kilauea, Hawaii, along with stones from Meteor Crater and the Petrified Forest, Arizona, and from important passes of California; a set of tiny Mayan terra-cotta heads and an obsidian sacrificial dagger; an old California Indian mortar stone, and rocks from the Indian cave Jeffers celebrated in the poem "Hands."[11]

Tor House was a geological event that occupied Jeffers's hands all his years. The original house (sitting room, bedroom, two attic rooms) was complete in 1919, the tower in 1924. Then he added a sociable dining room with literary quotations painted on its rafters—"Time and I against any two;" "*Carpent Poma Nepotes*" (Let the grandchildren gather the apples). One of the boulders in the wall has chiseled in it, "Hardy 1.11.28"—the date the stone

11 And hundreds more mementos chronicled in a pamphlet, *The Stones of Tor House*, by Donnan Jeffers, the poet's son. Jeffers Literary Properties, 1985.

12 *The Collected Poetry of Robinson Jeffers*, Vol. II, p. 131.

1928 - Jeffers stands in front of the fireplace in the dining room, which was completed in 1930. His unvarying routine was to write in the morning and labor with stone in the afternoon.

1968 - A view from the original house into the added-on dining room, with its fireplace just visible on the right. The shy poet made a room that welcomed guests from all over the world.

went in and writer Thomas Hardy died. In 1937 Jeffers began work on an adjoining house for his children and grandchildren. His son Donnan, having learned the craft, picked up the work in 1957 and continued it into the 1960s with a connecting wing between the two houses. In the 1990s Lee Jeffers, Robin's and Una's daughter-in-law, was still living in the new part of Tor House, with the old part open to the public by appointment.

Jeffers's art is in the house, and the house in his art. In 1932 he wrote "The Bed by the Window":

> I chose the bed down-stairs by the sea-window for a good death-bed
> When we built the house; it is ready waiting,
> Unused unless by some guest in a twelvemonth, who hardly suspects
> Its latter purpose. I often regard it,
> With neither dislike nor desire: rather with both, so equalled
> That they kill each other and crystalline interest
> Remains alone. We are safe to finish what we have to finish;

> And then it will sound rather like music
> When the patient daemon behind the screen of sea-rock and sky
> Thumps with his staff, and calls thrice: "Come, Jeffers."[12]

On a morning of rare snow thirty years later—January 20, 1962—Jeffers died in the bed by the window. Poets and duchesses know that doing a High Road building right is a labor of love measured in lifetimes.

CHAPTER 5

Magazine Architecture:
No Road

MOST BUILDINGS have neither High Road nor Low Road virtues. Instead they strenuously avoid any relationship whatever with time and what is considered its depredation. The very worst are famous new buildings, would-be famous buildings, imitation famous buildings, and imitation imitation buildings. Whatever the error is, it is catching.

I first came to this question—and to this book—through experience with a building whose fame, as usual, is the fame of its architect. I. M. Pei was considered by a 1991 poll of the American Institute of Architects to be "the most influential living American architect." Designer of the glass pyramid and renovation for the Louvre in Paris and of a spectacular skyscraper in Hong Kong, his reputation is worldwide. The MIT campus in Massachusetts considers itself lucky to have three buildings designed by Pei, who was once a student there. In 1986 his third MIT building, known informally as the Media Lab and formally as the Wiesner Building, had just opened when I joined the Media Lab for a season as a visiting scientist. What I have to say about the building does not reflect on the people or the activities of the Media Laboratory itself, which I enjoyed so much I wrote a book about them.[1]

It may have been my familiarity with MIT's homely, accommodating Building 20 just across the street that made the $45 million pretentiousness, ill-functionality, and non-adaptability of the Media Lab building so shocking to me. Here was a building purpose-built to house a diverse array of disciplines and people collaborating on deep research in fast-evolving computer and communication technologies. Consider in that light the building's dominant feature—its vast, sterile atrium. In many research

20 November 1990. Brand.

1990 - Blessed with not one but two echoing lobbies, the Wiesner Building's only cordial staircase links them. Foot traffic in all the working part of the building is forced to use depressing, hidden, concrete fire stairs. Researchers who are meant to collaborate across disciplines can go for weeks in the building without ever seeing each other casually. Desperate for more office and lab space, the Media Lab is considering tunnelling under the adjoining plaza.

IMPRESSIVE AND USELESS, the massive atrium in I. M. Pei's Wiesner Building at MIT hogs space, isolates and overwhelms people, and provides no amenities. It was meant to be a work of art, a continuation of the building's starkly Modernist skin (right) into the interior. The three balconies at upper left that look like truncated freeway exits serve no function whatever. The smoked-glass interior windows at upper right ensure that humans remain invisible in the space. The design may be alienating, but it's expensive.

20 November 1990. Brand.

1990 - Occupied in 1985, MIT's Wiesner Building houses art galleries on the ground floor and The Media Laboratory in the basement and top three floors. The design of more convivial research buildings is not a mystery—see p. 172 (Mathematical Sciences Research Institute), p. 176 (MIT's own Main Building), and p. 180 (Princeton's Molecular Biology Laboratory).

buildings a central atrium serves to bring people together with open stairways, casual meeting areas, and a shared entrance where everyone sees each other daily. The Media Lab's atrium cuts people off from each other. There are three widely separated entrances (each huge and glassy), three elevators, few stairs, and from nowhere can you see other humans in the five-story-high space. Where people might be visible, they are carefully obscured by internal windows of smoked glass.

The atrium uses up so much of the building that actual working office and lab space is severely limited, making growth and new programs nearly impossible and exacerbating academic turf battles from the first day. Nowhere in the whole building is there a place for casual meetings, except for a tiny, overused kitchen. Corridors are narrow and barren. Getting new cabling through the interior concrete walls—a necessity in such a laboratory—requires bringing in jackhammers. You can't even move office walls around, thanks to the overhead fluorescent lights being at a Pei-signature 45-degree angle to everything else.[2]

The Media Lab building, I discovered, is not unusually bad. Its badness is the norm in new buildings overdesigned by architects. How did architects come to be such an obstacle to adaptivity in

buildings? That's a central question not just for building users but for the architectural profession, which regards itself these years as being in crisis. Design professor C. Thomas Mitchell voices a common indictment:

> A range of observers of architecture are now suggesting that the field may be bankrupt, the profession itself impotent, and the methods inapplicable to contemporary design tasks. It is further suggested that collectively they are incapable of producing pleasant, livable, and humane environments, except perhaps occasionally and then only by chance.[3]

---

[1] Stewart Brand, *The Media Lab* (New York: Viking Penguin, 1987).

[2] The saga of the problem-filled creation of the building is told in *Artists and Architects Collaborate: Designing the Wiesner Building* (Cambridge: MIT Committee on the Visual Arts, 1985).

[3] C. Thomas Mitchell, *Redefining Designing* (New York: Van Nostrand Reinhold, 1993), p. 30.

At the risk of sounding like yet another architect-basher, let me take a stab at diagnosis.

Marvin Minsky, one of the founders of Artificial Intelligence, was gazing across the deserted Media Lab atrium with me one day. "The problem with architects," he rasped, "is they think they're artists, and they're not very competent."

Sure enough, if you look at the history of architecture as a profession, it was always around distinctions of "art" that architects distinguished themselves from mere "builders"—starting in the mid-19th century, when the profession emerged, and continuing to the present day.[4] "Art-and-Architecture" are always clumped together. What's wrong with that? Few modern artists approve of anything static. Artistic architects should be the most accepting of all of interactivity with their audience.

The problems of "art" as architectural aspiration come down to these:

- Art is proudly non-functional and impractical.

- Art reveres the new and despises the conventional.

- Architectural art sells at a distance.

Architect Peter Calthorpe maintains that many of the follies of his profession would vanish if architects simply decided that what they do is craft instead of art. The distinction is fundamental, according to folklorist Henry Glassie: "If a pleasure-giving function predominates, the artifact is called art; if a practical function predominates, it is called craft."[5] Craft is something useful made *with* artfulness, with close attention to detail. So should buildings be.

Art must be inherently radical, but buildings are inherently conservative. Art must experiment to do its job. Most experiments fail. Art costs extra. How much extra are you willing to pay to live in a failed experiment? Art flouts convention. Convention became conventional because it works. Aspiring to art means aspiring to a building that almost certainly cannot work, because the old good solutions are thrown away. The roof has a dramatic new look, and it leaks dramatically.

Art begets fashion; fashion means style; style is made of illusion (granite veneer pretending to be solid; facade columns pretending to hold up something); and illusion is no friend to function. The fashion game is fun for architects to play and diverting for the public to watch, but it's deadly for building users. When the height of fashion moves on, they're the ones left behind, stuck in a building that was designed to look good rather than work well, and now it doesn't even look good. They spend their day trapped in someone else's taste, which everyone now agrees is bad taste. Here, time becomes a problem for buildings. Fashion can only advance by punishing the no-longer-fashionable. Formerly stylish clothing you can throw or give away; a building goes on looking ever more out-of-it, decade after decade, until a new skin is

---

[4] The best source on this history is Andrew Saint, *The Image of the Architect* (New Haven and London: Yale, 1983). See Recommended Bibliography.

[5] Henry Glassie, "Folk Art," in *Material Culture Studies in America* (Nashville, TN: Am. Assoc. for State and Local History, 1982) ed. Thomas J. Schlereth, p. 126.

A standard line among potters is: "If it holds water, it's craft. If it leaks, it's art."

FAKE ASHLAR ON REAL ASHLAR. The illusion of perfectly-dressed stone was incised in stucco over very well-dressed genuine stone at the Classical Greek stadium (4th century B.C.) at Arycanda in modern Turkey. Through the ages, style-conscious building designers have preferred *faux* materials to the real thing. Early Modernist theory sneered at such fakery, but later Modernist architects were eager to clad their boxes with phoney veneers.

grafted on at great expense, and the cycle begins again—months of glory, years of shame.

A major culprit is architectural photography, according to a group of Architecture Department faculty I had lunch with at the University of California, Berkeley. Clare Cooper Marcus said it most clearly: "You get work through getting awards, and the award system is based on photographs. Not use. Not context. Just purely visual photographs taken before people start using the building." Tales were told of ambitious architects specifically designing their buildings to photograph well at the expense of performing well.

I heard something similar from San Francisco architect Herb McLaughlin: "Awards never reflect functionality. I remember serving on a jury one time and suggesting, 'Okay, we've winnowed this down to ten projects that we really like. Let's call the clients and see how they feel about the buildings, because I don't want to give an award to a building that doesn't work.' I was *hooted* down by my fellow architects." In London, architect

Frank Duffy fumed to me about "the curse of architectural photography, which is all about the wonderfully composed shot, the absolutely lifeless picture that takes time out of architecture—the photograph taken the day before move-in. That's what you get awards for, that's what you make a career based on. All those lovely but empty stills of uninhabited and uninhabitable spaces have squeezed more life out of architecture than perhaps any other single factor."

At a building preservation conference in Charleston, South Carolina, I chatted with an architecture student. Interested primarily in rehab and restoration work, she referred unflatteringly to the majority of her 450 fellow students at the Tulane University Architecture Department as "magazine architects." By which she meant image-driven and fad-driven architects, because architecture magazines probe no deeper than the look and style of the buildings they cover. They never interview clients or users. They never criticize buildings except, rarely, in terms of being bad art or off-trend. Articles consist primarily of stylized color photographs. Reports cover only new or newly renovated buildings, often in language that sounds like the "prismatic luminescence" school of wine writing. The subject is taste, not use; commercial success, not operational success. Architecture magazines are about what sells. They are advertising, cover to cover.

Architecture, being art, is supposed to have movements. In the customary sequence, a movement starts in relevance, loses it but prospers anyway for many years, then succumbs to the next cycle. Modernism began in the 1890s with the arrival of whole new *kinds* of buildings—factories, department stores, highrises. Form, stuck for precedent, followed function. Then that insight was aestheticized (and anesthetized) by the Bauhaus movement in the 1920s and 1930s, and its teachers migrated to America to avoid Hitler. So matters stood when all of America and Europe needed new buildings after the war—America to express its new prowess and prosperity, Europe to replace its devastated cities. Modernism got renamed the International Style ("because no one wanted to claim it," says Boston city planner Stephen Coyle), and suddenly

1993 - Ashlar, with its vertical and horizontal joints, is the most prestigious of masonry techniques. Arycanda is sited on the steep face of a mountain. This was a retaining wall behind the stadium seating. The stones are massive.

13 June 1993. Brand.

all the world's major cities began to look the same. Being minimalist on principle, Modernist buildings could be built cheap but showy. Less money for construction, more glory for the architect. Everybody won but the users.

In the late 1960s and early 1970s, architecture departments on a few campuses such as MIT, Berkeley, and London's Architectural Association fomented a revolution in response to the painfully evident failures of Modernism. Teachers began to talk of the need for "loose fit" in designing buildings, so that unexpected uses of the building could be accommodated.[6] "Post-Occupancy Evaluation" was invented and became a subdiscipline. Chris Alexander and colleagues came up with their "pattern language" of design elements that wear well. The rapid growth of services such as air-conditioning led to more systemic study of buildings and their uses. Conserving energy became a source of design innovation. Even the psychology of building use was explored. Then it all faded. What happened?

Art fought back. In 1966 the Museum of Modern Art had published a book by architect Robert Venturi titled *Complexity and Contradiction in Architecture*. It attacked the puritanical simplicity of Modernism—"I am for messy vitality over obvious unity." In 1972 Venturi aesthetically interpreted a commercial event—the gaudy success in Las Vegas of what he called "decorated sheds."[7] The insight was perfect. Now architects could celebrate the corner they had been backed into by the fragmentation of the building trades, and be exterior decorators solely and proudly. I don't fault Venturi; I admire his buildings and writings. It was the leadership of the profession that took the easy path away from complex responsibility and back to airy debate about style.

By the 1980s, "Post-Modernism" had swept the architecture departments and the rising partnerships. "Historicism"—casual looting of historic styles—took off, and Palladian half-circles, triangular gable ends and toy-like cheery colors bedizened the land. To their credit, Post-Modern buildings were general

purpose and adaptive inside—decorated *sheds*.

But their appeal was shallow. Architect Peter Calthorpe says it's as if architecture suffered a stroke after World War II, lost its language and intelligence, and couldn't articulate real buildings any more, so it just made boxes. With Post-Modernism, the profession was finally beginning to manage a few words again, but it still couldn't handle sentences or paragraphs. Instead of settling down to real homework, the Post-Modern movement increasingly fragmented into shards of self-consciousness. Obsession with such distractions as French literary criticism culminated with a suicide dive into blatantly "postfunctional" Deconstructionism.

By the early 1990s architecture was adrift—toying with Neo-Modernism, Late Modernism, Second Modernity, Neo-Classicism, Neo-Rationalism, and other signs of being open to renewed investigations of relevance. In an essay titled *For an Architecture of Reality*, architect Michael Benedikt reminded his colleagues, "We count upon our buildings to form the stable matrix of our lives, to protect us, to stand up to us, to give us addresses, and not to be made of mirrors."[8]

A building's exterior is a strange thing to concentrate on anyway. All that effort goes into impressing the wrong people—passers-by

From Repton's beautiful "Red Book" at the British Architectural Library, Royal Institute of British Architects (RIBA). Neg. no. 0272.

"FACADECTOMY" is no recent phenomenon. Eighteenth-century architect and landscape gardener Humphrey Repton (1752-1818) offered the service to his clients with clever watercolor renderings. The viewer untabbed and lowered the flap to reveal the new look in the old setting.

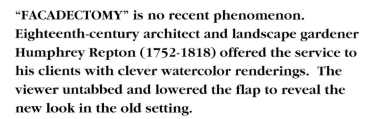

instead of the people who use the building. Only if there is a heavily trafficked courtyard or garden do the building dwellers notice the exterior at all after the first few days. Most often they don't even enter by way of the facade and big lobby; they come in by the garage door. And yet, ever since the Renaissance, "the history of architecture is the history of facades." It is a massive misdirection of money and design effort, considering how badly

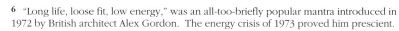

6 "Long life, loose fit, low energy," was an all-too-briefly popular mantra introduced in 1972 by British architect Alex Gordon. The energy crisis of 1973 proved him prescient.

7 Robert Venturi, et al., *Learning from Las Vegas : The Forgotten Symbolism of Architectural Form* (Cambridge: MIT Press, 1972, 1977).

8 Michael Benedikt, *For an Architecture of Reality* (New York: Lumen, 1987), p. 14.

9 Francis Duffy, *The Changing Workplace* (London: Phaidon, 1992), p. 232.

buildings need their fundamentals taken care of. Chris Alexander is vehement: "Our present attitude is all reversed. What you have is extremely *in*expensive structure and all this glitz on the surface. The structure rots after thirty years, and the glitz is so expensive that you daren't even fuck with it."

Architects got themselves stuck in the skin trade. Frank Duffy observes, "The only area of architectural discretion in artistic or financial terms is the skin. The architectural imagination has allowed itself to be well and truly marginalized."[9] It happened because architects offered themselves as providers of instant solutions, and only the look of a building gives instant gratification. When the space planning doesn't work out and needs improvement, or the structure indeed rots, where's the architect? Long gone.

There is little public accounting. All too typical was one of Britain's most celebrated architects, the late Sir James Stirling, recipient of a Gold Medal from the Royal Institute of British Architects and a Pritzker Prize as well as a knighthood. The honors kept coming despite widely reported disasters with his buildings. His Engineering School at the University of Leicester (1966) dropped its roof tiles and leaked so badly it had to be reconstructed at huge expense. (Architectural critics spoke approvingly of Stirling's "mannerist ambiguity" used to "willfully repulsive effect.") His History Faculty building at Cambridge (1967) leaked torrentially, also dropped its roof tiles dangerously, and faced the wrong way, so the sun cooked its contents – people and books. It was so noisy and unusable that the faculty almost voted to demolish it. It also was rebuilt at huge expense. His low-cost housing complex at Southgate (1977) was so loathed by its occupants that it *was* demolished, just twelve years after it was built. Reputations based on exterior originality miss everything important. They have nothing to do with what buildings do all day and almost nothing to do with what architects do all day.

Does the building manage to keep the rain out? That's a core issue seldom mentioned in the magazines but incessantly mentioned by building users, usually through clenched teeth. They can't believe it when their expensive new building, perhaps by a famous architect, crafted with up-to-the-minute high-tech materials, *leaks*. The flat roof leaks, the parapets leak, the Modernist right angle between roof and wall leaks, the numerous service penetrations through the roof leak; the wall itself, made of a single layer of snazzy new material and without benefit of roof overhang, leaks. In the 1980s, 80 percent of the ever-growing postconstruction claims against architects were for leaks.

Architects' reputations should rot if their buildings can't handle rain. Frank Lloyd Wright was chosen by a poll of the American Institute of Architects as "the greatest American architect of all time." They all knew his damp secret:

> Leaks are a given in any Wright house. Indeed, the architect has been notorious not only for his leaks but for his flippant dismissals of client complaints. He reportedly asserted that, "If the roof doesn't leak, the architect hasn't been creative enough." His stock response to clients who complained of leaking roofs was, "That's how you can tell it's a roof."[11]

Wright's late-in-life triumph, Fallingwater in Pennsylvania, celebrated by the AIA poll as "the best all-time work of American architecture," lives up to its name with a plague of leaks; they have marred the windows and stone walls and deteriorated the structural concrete. To its original owner, Fallingwater was known as "Rising Mildew," a "seven-bucket building." It is indeed a gorgeous and influential house, but unlivable. For its leaks there can be no excuse.

Wright is an interesting study of a superstar architect having both right and wrong influence. "All Architecture, worthy the name," he decreed in 1910, "will, henceforward, more and more be organic."[12] So inspired by Viollet-le-Duc and Louis Sullivan, he inspired countless others (including young me) toward an organic approach to architecture. At the same time, the very pomposity of his decrees helped inflame a fatal egotism in generations of architects, and his most famous buildings belie his organic ideal. They were so totally designed—down to the screwheads all being aligned horizontally to match his prairie line—that they cannot be changed. To live in one of his houses is to be the curator of a Frank Lloyd Wright museum; don't even think of altering anything the master touched. They are not living homes but petrified art, organic only in idea, stillborn.

A large part of Wright's problem with exasperated clients (and with rain) can be traced to his impatience with the traditional rectangular shape of buildings. "I have, lifelong, been fighting the pull of the specious old *box*," he told the AIA in 1952,[13] and too many listened. Architects love exotic building shapes. Wright

---

11 Judith Donahue, "Fixing Fallingwater's Flaws," *Architecture* (Nov., 1989), p. 100. Someone should have replied to Wright, "That's how you can tell it's a failure."

12 *Frank Lloyd Wright: Writings and Buildings*, Edgar Kaufmann and Ben Raeburn, eds., (New York: Meridian, 1960), p. 90.

13 Above title, p. 289.

14 George Oakes in Lloyd Kahn's *Refried Domes*, p. 1, (1989; $7 postpaid from Shelter Publications, P.O. Box 279, Bolinas, CA 94924). Lloyd Kahn's enthusiastic earlier book, *Domebook II* (1972), sold 175,000 copies. By the late 1970s, his grim experience with domes converted him from a proponent and builder to a critic. He tore down his own dome home and replaced it with a conventional structure.

April 1988. © Lloyd Kahn.

1988 - Though it was competently built by Bill Woods of Dyna Domes, Phoenix, soundness and sheer flash were not enough to keep so unwieldy a space occupied.

ABANDONED DOME. A combination restaurant, bar, and real estate office, this geodesic dome stands empty by the highway near Yucca, Arizona. How would one add to it? The difficulty of dealing with a curved exterior is apparent in the stairs/fire escape grappling to reach three floors. Inside is even more awkward. It is an amazing structure to look at, though.

was always experimenting with hexagonal, triangular, and round shapes in his structures. They were impressive, but they fought the hand of dwellers as well as builders. Wright didn't care. I. M. Pei has shown similar obliviousness in such buildings as the highly admired, severely triangular East Wing of the National Gallery in Washington, DC. Every space in the building is a pain to work with. Even the shelves in the voluminous library end in acute or obtuse angles that won't hold a book; special, space-wasting blocks, different for each shelf end, had to be fashioned. The crowning indulgence was a triangular elevator, gratuitously awkward, astronomically expensive.

As for domes, fancied by architects through the ages, the findings are now in, based on an entire generation's experience with Buckminster Fuller geodesic domes in the 1970s. They were much touted in the architecture magazines of the period. As a

major propagandist for Fuller domes in my *Whole Earth Catalogs*, I can report with mixed chagrin and glee that they were a massive, total failure. Count the ways…

Domes leaked, always. The angles between the facets could never be sealed successfully. If you gave up and tried to shingle the whole damn thing—dangerous process, ugly result—the nearly horizontal shingles on top still took in water. The inside was basically one big room, impossible to subdivide, with too much space wasted up high. The shape made it a whispering gallery that broadcast private sounds to everyone in the dome. Construction was a nightmare because *everything* was non-standard—"Contractors who have worked on domes all swear that they'll never do another."[14] Even the vaunted advantage of saving on materials with a dome didn't work out, because cutting triangles and pentagons from rectangular sheets of plywood left

THE OPPOSITE OF A DOME in every respect is a plain old rectangular Cape Cod house. Nearly hidden behind later additions, the original house here (1825) is at the left rear. Right angles, flat walls, and simple roofs made it easy to add a tower, a wraparound porch, a two-story lean-to, a kitchen ell (with narrow stove chimney), and a one-story lean-to connecting to the barn at the right. The add-on look of "cascading roofs" is currently much imitated in brand new houses. (Too bad. It makes them expensive and hard to add on to. Leaping at the look prevents the reality.)

ca. 1890 - This amalgam near Truro, Massachusetts, was recently reduced by a Cape Cod purist to just the original house and an ell. What is added to a rectangular building can be subtracted fairly gracefully.

enormous waste. Insulating was a huge hassle. Doors and windows weakened the structure, and *they* leaked because of shape and angle problems.

Worst of all, domes couldn't grow or adapt. Redefining space inside was difficult, adding anything to the outside nearly impossible—a cut-and-try process of matching compound angles and curves. When my generation outgrew the domes, we simply left them empty, like hatchlings leaving their eggshells.

We didn't know it at the time, but the Fuller dome craze was a reprise of an earlier American architectural folly. In 1849 a phrenologist named Orson Fowler published a book titled *A Home for All* ,[15] which lauded the benefits of octagon-shaped houses. A minor rage for building the things swept America. Only a few survive. Like domes, they proved awkward to subdivide or add to, and the fad was self-extinguishing. Fowler

eventually moved out of his own octagon. When he built again in 1877, it was a conventional house. (Neither Fowler nor Fuller, it must be noted, were architects.)

Architects perpetually seeking a radical new look should study the lessons of domes. Dome apostate Lloyd Kahn rediscovered verticality: "What's good about 90-degree walls: they don't catch dust, rain doesn't sit on them, easy to add to; gravity, not tension, holds them in place. It's easy to build in counters, shelves, arrange furniture, bathtubs, beds. *We* are 90 degrees to the earth."[16] Occupants are always intuitively oriented in rectangular buildings. (Orientation itself is a right-angle concept, the cardinal directions being at 90 degrees.) Since buildings inevitably grow, it is handy to have the five directions (including up) that rectangular buildings offer for expansion. Right-angled shapes nest and tile with each other universally, so tables fit into corners, and clothes into closets, and buildings into city lots, and lots into city blocks. Humanity's first cities in Sumer had rectangular street grids. Tribal cultures the world over have beautiful round dwellings, but the shape of civilization is rectangular.

The "specious old box" is old because it is profoundly adaptive.

But it's not just because of impractical designs that architects are only involved in 5 percent of new buildings. The architect is being marginalized by the pathological fragmentation of the building professions and trades. Is the architect a generalist, a maker of buildings, or a specialist, a mere artist? That question haunts the profession, because students are attracted to architecture as a wondrous calling for great souls guiding huge projects with all-embracing talent and skill. After graduation they encounter a tawdry reality—architect as deskilled and disempowered minor player who is increasingly left out entirely.

A standard commercial building project is set in motion by a speculator (who may not plan to be the landlord) contracting with an architectural office for a building design. The design goes through a gauntlet of permits and emerges distorted, if at all. At this point the work is handed over to a battery of engineers— structural, service, "value," etc.—who have been trained to a completely different discipline than the architect and who are scrupulously shielded from any skill or interest in aesthetic design. Then responsibility shifts to a general contractor, with the architect now reduced to an observational rather than supervisory role. The contractor passes 80 percent of the work to subcontractors.

---

[15] Still in print. Orson Fowler, *The Octagon House: A Home for All* (New York: Dover, 1853, 1973).

[16] *Refried Domes*, p. 7.

[17] A bracing read for American and European professionals is: Sidney Levy, *Japanese Construction* (New York: Van Nostrand Reinhold, 1990). *The Economist* reports, "In Japan it is common for the architect's drawings to be altered radically on the building site. All such sites have field offices, known as *genba*, where architects and builders work together. This helps them to collaborate on the unforeseen problem of detail, the sudden glitch." "That Certain Japanese Lightness," *The Economist* (22 Aug. 1992), p. 75.

They are often the ones with the cutting-edge technical skills, but they are too far downstream to affect design. Once the building is finished, it is turned over to facilities managers who will actually run the building. They of course have had no hand in its design. The speculator sells to a landlord, who rents to tenants, whose sole design function is to pay retroactively for the whole mess.

The result? The British editor of this book, Ravi Mirchandani, has a tale identical to what I encountered in a dozen other publishing offices in London and New York: "Penguin's building is too hot in the winter and too cold in the summer. The air is stale; everyone has colds that are endless. No one likes the building, and it's only five or six years old. Nobody knows the designer's name—just some team who retreated into the shadows."

The process has evolved in part to disperse responsibility and foil lawsuits. It fails in that respect too. Lawsuits increase yearly. So far, two partial solutions to the fragmentation problem have emerged, one characteristically Japanese, one pragmatically American.

Japan has turned a cohesive design-build approach into mega-success with a half dozen multi-billion-dollar firms like Kajima and Obayashi, capable of building whole cities. Employing Japanese traditions of collaboration and long-term commitment, these companies offer one-stop-shop, cradle-to-grave service. From one closely coordinated group you get financing, site selection, building design, construction, interior design, furnishings, facilities management, and maintenance. The designers and facilities managers have to listen to each other, because they work for the same company. The client (often the tenant) can complain to one person in the system and expect timely correction. Japanese custom discourages lawsuits and encourages reality-based accommodation. Architecture, treated as a craft rather than an art, is taught in engineering departments. The Japanese design-build methodology has developed to such efficiency that some highrises have construction begun on their base before the top is complete-ly designed—"just-in-time" design.[17] Research & Development gets ten times what American firms spend. In the international market these companies are often murdering American and European competition. And their buildings don't leak.

The American solution is much more combative. Here the illuminating book is Joel Garreau's *Edge City*, which describes the teeming new commercial complexes arising around the periphery of cities:

> The height, shape, size, density, orientation, and materials of most buildings are largely determined by the formulaic economics of the Deal. It was stunning how completely it was the developers who turned out to be our master city builders. The developers were the ones who envisioned the projects, acquired the land, exercised the planning, got the money, hired the architects if there were any, lined up the builders, and managed the project to completion. Often, it was the developers who continued to manage and maintain the buildings afterward…. Architects were lucky if they got to choose the skin of the building.[18]

The combat between architects and developers was decided some while ago. Architects are a minor ornament of the culture. Developers build America. There are countless departments and schools of architecture, but the only training for developers is deep within a few business schools. Most development is simply done by whoever has the money and the nerve and the ability to make people collaborate. Because developers are truly driven by the market, there is a rowdy but honest market feedback in the system. What doesn't work, they blithely tear down or remodel until it does work. Architects are too lofty to study what is going on until a peer like Robert Venturi declares, "Learn from Las Vegas;" and "Main Street is almost all right;" and "Developers build for markets rather than for Man and probably do less harm than authoritarian architects would do if they had the developers' power."[19]

Fragmentation envelops the customer as well. Clients often are no better at representing building users than architects are. Usually a building is so large and complex an undertaking that the "stakeholders" are too diverse, scattered, and at odds to agree on much of anything. Architects complain justifiably that client organizations typically bump "facilities" decisions and supervision down to about the third layer of management, where decisiveness and clarity are scarce. Sometimes the company's single most knowledgeable person about what is happening with the projected building is the lawyer hired to supervise all the multidirectional negotiations and agreements. Major design decisions wind up being made semi-randomly by the lawyer because no one is crisply in charge.

Another misdirection comes from the way architects and contractors are paid. The standard fee for an architect is around 10 percent of the building cost (6-7 percent for a commercial building). That's about right for the amount of work it takes, but it introduces a pernicious incentive. The client asks, "What if we added another bed and bath over here?" The architect says, "It might come to $50,000 extra"—thinking, *another $5,000 for me.* An unethical architect would remark, "It could be very nice on that side of the house," instead of advising, "You really don't need it, and you could add it later, especially if we design this wall here with that in mind." The percent approach is a conflict of interest for the architect; it encourages buildings that try to be too perfect and too large too soon. Better would be to agree with the architect in advance on a set budget and set fee for the building, with a significant bonus for the architect if the project stays within the budget.

However, an unscrupulous contractor can undermine that approach in a hurry. Construction costs get really corrupt in the area of "claims" arising from "change orders." Changes that come up during construction are billed punitively by the contractor. Frank Duffy is savage on the subject: "It's a racket. It's disgraceful. If you are a British contractor, the biggest department you will have is called 'Claims,' where modifications are costed and invoiced. You will have your best intelligence working in Claims.

It is a major industry, more important than building buildings. It is all based on the idea that only the client changes his mind. The building industry is a timeless and perfect automaton that exists only to carry out specific instructions. It doesn't have a mind of its own. In Claims the boundary between the timeless and the timeful is being negotiated—that's why it's such a hot issue. It's seismic. It leads to agony for the client and money for the contractor. It's very destructive; that's why it's so evil. It's also evil because it's stupid, it denies intelligence."

What is punished by claims is any kind of adaptivity during construction—exactly the time when you most want it, because the building is incomplete enough to fix and improve easily. Also that's when you can take advantage of the skills and intelligence of the artisans to fine-tune design solutions on the site.

The simultaneous seizing of power and shedding of responsibility by contractors puts the onus on architects to anticipate perfectly all of a building's needs. Nothing is left to the builders, to the client, or to actual usage. But if architects are now in a bind with that, they asked for it. They sought total control and promised total prescience. Frank Duffy: "Inherent to the Modern movement is the German idea of *gestalt*—totality. It's Bauhaus. It's a terribly powerful word that was interpreted by architects as the power to determine every detail of the building. And you cannot touch anything once it's there." Architects of this persuasion want your light switches and toilet stalls, desks and fire stairs to reflect the same pervasive aesthetic as the roof line and lobby. If your life is out of synch with that, too bad.

Gestalt design defies what should have been learned from the history-of-architecture courses (which were optional for architecture students in the Modernist years but now are required again). The most admired of old buildings, such as the Gothic palazzos of Venice, are time-drenched. The republic that lasted 800 years celebrated duration in its buildings by swirling together over time a kaleidoscope of periods and cultural styles all patched together in layers of mismatched fragments—an aesthetic practice that may be appropriate to our own multicultural age.

Instead of steady accumulation, the business of contemporary architecture is dominated by two instants in time. One is the moment of go-ahead, when the architect's reputation and the beguiling qualities of the renderings and model of the building-to-be overwhelm the client's resistance. The other is the moment of hand-over, when the building shifts from the responsibility of the builders to the responsibility of the owner, and occupancy begins. These are necessary moments to make coordination possible, but they distort everything. Each instant is a massive barrier to learning.

The effort is to make everything perfect and final for each of these opening nights. The finished-looking model and visually obsessive renderings dominate the let's-do-it meeting, so that shallow guesses are frozen as deep decisions. All the design intelligence gets forced to the earliest part of the building process, when everyone knows the least about what is really needed. "A lot of the time now, you see buildings that look *exactly* like their models," one model maker told me. "That's when you know you're in trouble."

Chris Alexander likes to make on-site adjustment to a building as it's being constructed. "Architects are supposed to be good visualizers, and we are," he says, "but still, most of the time we're wrong. Even when you build the things yourself and you're doing *good*, you're still making nine mistakes for every success. So you take the time to correct them. The more at each stage you can approach being able to experience the contemplated reality, the more it will give you feedback and you'll be able to intelligently develop it."

18 Joel Garreau, *Edge City* (New York: Doubleday, 1991), p. 326. See Recommended Bibliography.

19 Robert Venturi et al., *Learning From Las Vegas* (Cambridge: MIT, 1972), p. 106. Quoted in C. Thomas Mitchell, *Redefining Designing* (New York: Van Nostrand Reinhold, 1993), p. 17.

ca. 1968 - Four matched but different brick buildings from the 1840s and 1850s graced West 11th Street in Greenwich Village.

1970 - When the Weathermen's bomb factory exploded, it took several activist lives and demolished the house. Brick party walls protected the neighbors.

1980 - Like a New York wisecrack—cynical, sentimental, and slick—Hardy's design of the new house at Number 18 heightens its swiveled reference to the 1970 explosion by blending in all the other building details with the neighbors even more than the original house did..

URBANE MEMORY. On March 6, 1970, political activists (the Weathermen) accidently blew up themselves and an 1840s row house in New York. Architect Hugh Hardy designed a replacement that fit in politely with the neighbor buildings but also acknowledged what had happened on the site.

The time to correct mistakes is not available, shout the architects, the contractors, the bankers, and the clients. Right, groans Alexander, and that's why most buildings are crappy. "There is real misunderstanding about whether buildings are something dynamic or something static. The architect has such a narrow niche. Anything different from the idea that you make a set of drawings and someone else builds the thing is incredibly threatening. People get just absolutely freaked out. I think it's because it raises specters about contracts." Matisse Enzer, a contractor who has worked with Alexander, agrees: "Architects think of a building as a complete *thing*, while builders think of it and know it as a *sequence*—hole, then foundation, framing, roof, etc. The separation of design from making has resulted in a built environment that has no 'flow' to it—you simply cannot design an improvisation or an adaptation. It's dead."

Then comes the barrier of occupancy. A kind of frenzy grips the job site as contractor, client, and architect grind their way down the "punch list" that defines all the details to be finished before the job is done. The heroic joys of framing are long forgotten; now the finishing drags on and on, with querulous subcontractors, gaps between responsibilities, and scheduling hassles that make delays avalanche. Tempers snap, lawyers circle, and everyone just wants the nightmare to end. It ends in exhaustion. The building's users, makers, and designers want nothing to do with each other ever again.

The race for finality undermines the whole process. In reality, finishing is never finished, but the building is designed and constructed with fiendish thoroughness to deny that. The occupants are supposed to march in and gratefully do exactly what it was declared they would do two years before, during the design stage, and the building will punish them if they don't.

Landscape architects don't think that way. They know that what they design has its own life, and they plan accordingly. One of the greatest was "Capability" Brown, whose 18th century work in England is lauded by Deborah Devonshire, based on her experience with Brown's park at Chatsworth: "Brown planted the wood in wedge-shaped compartments, so when one section is mature and ready to be felled, another is growing and the line remains unbroken…. Brown's gift of foresight and his ability to picture how the scenery would look 200 years after he finished his work is evident in parks in nearly every county in England."[20]

Neither do space planners think about finality. Frank Duffy recalls the experience that turned him away from standard architecture in the late 1960s in New York City: "The space planners—sort of sub-architects—got a million things right that we hadn't. They understood time. They worked for the IBMs and the Salomon Brothers, who were always changing, and they invented a design service which was related to change in a totally unglamorous, very New Yorky, pragmatic sort of way. Few architects got involved; it came out of the commercial-interiors world. Philip Johnson, I remember, called them 'hacks'—no significance at all. But they got it right. That's where I learnt the insight. I was a consultant to them. I did a post-occupancy evaluation for Arthur Andersen in New York in 1968, which must have been one of the first POEs ever done."

The inane but now standard term "post-occupancy evaluation" (POE) shows what a divisive watershed the moment of occupancy is. One of architecture's most adaptive devices is misnamed by the traumatic instant of letting users into a building. Why not just "use-evaluation," since that's what it is?

POEs are a shockingly direct procedure for judging how well a building works by formally surveying the occupants, especially the people who clean, service, or repair the building and know its failures all too well. Trained observers also watch and photograph how the building is used, comparing what actually is happening against what was intended. POEs began as recently as 1967, with a study of highrise student dormitories at the University of California, Berkeley. The study showed that large, showy lounges and recreation rooms were seldom used, that students were frantic for quiet and privacy to study in, that student rooms and desks were way too small, and that the rooms could not even be decorated by their reluctant occupants. Student were fleeing the dorms for any kind of private apartment or shared house in Berkeley. The design of dormitories improved as a result of the study.[21]

Herbert McLaughlin, head of KMD, San Francisco's largest architectural firm, has thought about how POEs could work optimally: "We believe you should go back three times. You should do it with the people who are going to use the building six weeks before it opens, to record their expectations. That gives you a very interesting base. You should go back within the first six months, when they're still fresh on the place and really feeling all the uneasy elements. Then you should go back about two years later, after they've accommodated themselves to the building. It would be wonderful to do a fourth one maybe ten years later, because by then the world has changed. Has your building been able to accommodate that change?" Such a sequence would measure expectation, initial frustration and delight, adjustment, and long-term adaptivity. It's never been done.

The few POEs that do get done are mostly hidden. *Architecture* magazine explains why:

> Architects generally have been slow to join the evaluation bandwagon. More often than not, the pull to conduct evaluations has come from client organizations, not from the architects themselves. Many architects in the past have regarded POE as negative feedback…. Organizations willing to pay for POEs tend to develop buildings of the same type on a regular basis and are interested in perfecting the end result…. Much of the private evaluation work done is

20 Deborah Devonshire, *The Estate* (London: Macmillan, 1990), p. 149.

21 Sim Van der Ryn and Murray Silverstein, *Dorms at Berkeley* (Berkeley: UC Center for Planning and Research, 1967).

Friday mid-afternoon, 16 August 1991. Brand.

1991 - Excessive high-rise glass, as in New Zealand House, is a many-leveled error—it punishes the occupants perpetually, it advertises its folly to the public, and it is prohibitively expensive to correct. The building's curtains track the sun like a negative sunflower.

"THE SINS OF THE ARCHITECT are permanent sins," wrote Frank Lloyd Wright. London's sore-thumb highrise, New Zealand House, has glass curtain walls that force the occupants to pull reflective curtains whenever the sun comes out, or bake. So much for the view.

ANTICIPATING TIME (right), architect Arthur Beresford Pite (1861-1934) rendered his new mission hall as if it was already old, to blend in with London's East End slums around it. The practice is so rare as to be collectible.

proprietary. Organizations such as Marriot, McDonald's, IBM, Steelcase, and Red Lobster all perform POEs in some fashion, but the results rarely are published.[22]

I recall asking one architect what he learned from his earlier buildings. "Oh, you never go back!" he exclaimed, "It's too discouraging." That answer inspired this chapter. In a remarkable study of fifty-eight new business buildings near London, researchers found that in only one case in ten did the architect ever return to the building—and then with no interest in evaluation. The facilities managers interviewed for the study had universally acid views about the architects. One said, "Their primary interest is in aesthetics rather than practicality; it is a ludicrous situation. It's no good if the building looks nice and doesn't work." Another: "They design it and move on to the next one. They're paid their fee and don't want to know."[23]

Frank Duffy regrets what has happened to post-occupancy evaluation since he started with it in 1968. "It's got somewhat trapped in the academic field, I'm afraid," he said. Architects seldom get around to it. "The fundamental reason is: the difficulty of putting buildings up is so great, and the pressures of getting it right on the night are so enormous, that squeezes out concern for the user and it squeezes out concern for time. I think architects have a tremendous responsibility for change here, because we ought to be the gateway between the construction and the consumer."

His view is supported by a growing practice in the industry of hiring "project managers" instead of architects to run the construction process. According to one book, "The construction management sector did not exist in 1967, but today its dollar volume rivals that of architectural firms."[24] This is a classic case of an overly self-protective profession being steadily displaced by paraprofessionals. Architects must widen their skill sets or continue dwindling into resented irrelevance.

ca. 1891 - Pite's mission was on Old Montague Street, Whitechapel. None of the buildings survived redevelopment.

our boys went over there and studied architecture, a professional architect was a man with a job and a salary, usually assigned to one great building. Architects now have all this schooling. They have to run a business when they get out of school or starve, and it's a starvation business to be in, because the overhead is consuming, and sometimes the wave just gets them. They spend the rest of their careers peddling what they learned in school. Most architects I know are hustling their tails to survive. They don't have the time to enrich themselves, to learn. They're too busy to grow. I think a creative person has to constantly grow, or forget it."

How close to the edge are architects? In the recession of the early 1970s, half of New York's architectural firms went out of business.[25] The combination of recession and a real-estate collapse in the early 1990s was even more brutal nationwide. It is in the nature of hustling for a living that the hustler must always focus on the *next* job, giving just enough attention to current jobs to get by, and no attention at all to past jobs. Two famed California architects, Bernard Maybeck and Charles Greene (of Greene and Greene), overcame the discontinuity problem by performing constant experiments on their own homes. Those houses became showcases, not of some finished theory, but of lifelong never-finished learning. A walk through their houses was said to be like a walk through the history of their creative development. The instance is noteworthy because it is rare.

One corrective has emerged. In the 1980s, malpractice lawsuits against architects surpassed those against doctors. Architects are learning painfully in court what they did not learn in school or from their buildings about leaky roofs, feeble structure, shoddy materials, poor detailing, inept window design. The same harsh lessons are driven in by the providers of their expensive "errors and omissions" insurance. Distinct improvement in the profession has resulted. So have some evasive techniques, such as grouping the designers of a public building together in a shallow corporation that conveniently dies when the lawsuits begin.

It may be difficult. Architects and their profession are constitutionally resistant to learning. The restoration specialist (and architect) William Seale gives historical perspective: "I have a theory about architects. In France in the 19th century, when all

[22] Elena Marcheso Moreno, "The Many Uses of Postoccupancy Evaluation," *Architecture* (Apr. 1989), pp. 119-121.

23 *The Occupier's View: Business Space in the '90s.* This 1990 study is the best public, general-purpose POE I have seen, rare and priceless (if pricy). Done by a surveyors' (real estate) firm called Vail Williams, it costs £50 from Vail Williams, 43 High Street, Fareham, Hampshire, PO16 7BQ, England. See Recommended Bibliography.

24 Preiser, Rabinowitz, and White, *Post-Occupancy Evaluation* (New York, Van Nostrand Reinhold, 1988), p. 28. Architects are increasingly regarded as out-of-it tyrants. "Large organisations are now convinced of [the] link between adaptability and survival, often preferring to commission new buildings with a project manager rather than an architect in charge. In the traditional, vertical design team, the services and structural engineers have no real chance of equality with the generalist architect. The operational stakes are now acknowledged by the more sophisticated companies to be too high to allow non-specialists in these areas to take overall control." Steelcase Stafor, *The Responsive Office* (Streatley-on-Thames: Polymath, 1990), p. 73.

[25] Judith Blau, *Architects and Firms* (Cambridge: MIT, 1984), cited in Dana Cuff, "Fragmented Dreams, Flexible Practices," *Architecture* (May 1992), p. 80.

68

1903 - The house at 400 Arroyo View Drive in Pasadena was designed by Greene & Greene as a simple two-story rental for a real-estate speculator.

GOING BACK to an earlier building can be one of the great treats, both for architect and client. As their Craftsman style and technique improved from year to year, California's Charles and Henry Greene were often invited back to upgrade prior houses. They learned what features wore well with time and what turned out to be mistakes, and they got to fix them. This bungalow in Pasadena went from a low-budget rental unit to a prime home.

1908 - James W. Neill bought the house for a residence and invited the Greenes to exercise their growing craft. They nailed shingles over the clapboard siding, expanded the living room to include the side porch, and added a front porch, a brick-and-boulder buttressed terrace, and a pergola over the drive.

Legal action is a hugely inefficient method of failure analysis. To decrease costs and vulnerability, architects—individually and as a profession—would do well to explore other means. Something very specific of this sort has been set up in New York State, where "architect Richard Bodane and his colleagues in the Office of General Services…have developed a database on roof design and performance. The database tracks a growing number of the state's 10,000 buildings, recording variables of location, design conditions, specified components, testing results, and the history of problems and their solutions. By correlating design information with performance problems, the architects identify patterns of success and failure."[26] How interesting it would be to go even deeper in the analysis: what were the organizational patterns associated with the roof successes and failures? Which arrangements can detect crucial errors and correct them, and which cannot? The answer to that question could affect all of architecture.

A cruel but pointed project would be a systematic study of the failings of the buildings that house university architecture departments. Invariably designed and built with great fanfare, as a class they are perhaps the most loathed of all academic buildings. From Paul Rudolph's infamously brutalist Art and Architecture Building at Yale to the harsh Wurster Hall at Berkeley, the buildings are exciting and unworkable. Are they unworkable *because* they are exciting? Do these theory-based "statements" and sculptural experiments necessarily and always impede daily function? Also: what do the students learn from these buildings—to avoid the mistakes, or to imitate them? After all, the buildings got built and got noticed, and those are the main criteria of success for an architect as the profession stands today.

The field of architecture takes its counsel from two main sources—architecture schools and architecture magazines—both

[26] B. J. Novitski, "Roofing Systems Software," *Architecture* (Feb. 1992), p. 102.

ca. 1922 - The original, hip-roofed part (right) of "The Ark" was designed in 1906 by the Architecture Department's founding chairman, John Galen Howard. Each new addition retained the original's homeyness and its Berkeley Shingle style. The building grew around a courtyard, faced on one side by a long, sunny corridor connecting all the studios, where teachers felt comfortable sticking their heads in to see what was up. Built originally as a temporary building, it now is on the National Register of Historic Places.

1993 - Completed in 1964, the School of Environmental Design was designed by Joseph Esherick, Vernon DeMars, and Donald Olsen to serve 1400 people (60 faculty). Professor Van der Ryn recalls, "William Wurster commissioned three faculty with entirely different philosophies to design the building. He thought it would be interesting to put them together and see what happened. What happened was an incredible struggle and a building that I don't think is even as good as the most mediocre work of any one of them. There are very few seminar rooms. The architecture studios are completely chaotic. Students walk back to where they've set up, and their desk is gone. The mechanical systems are ridiculous—in the summer in the south rooms the heat is going, and in the winter it's off. Luckily the windows do open."

THE LOVED AND THE UNLOVED. In 1964 the Architecture Department of the University of California, Berkeley, moved from an old building (above) to a new (right)—from one that had grown with the department for 58 years to one that was specially designed for it. The old one, known as "The Ark," is described as "one of the most revered buildings on the campus." As for the new one, Wurster Hall, "It's a real brute of a building," says architecture professor Sim Van der Ryn, who taught in both buildings. He explains, "I prefer a one-story responsive wood building to a ten-story extremely inflexible concrete one. When you go from a horizontal plan to a vertical one, immediately you develop a kind of hierarchy, and there's less communication. Sometimes I teach for a whole semester now and people say, 'I thought you were on leave.'"

Like a New England connected farm, "The Ark" grew around a south-facing courtyard, adding studios, an exhibit hall, a lecture hall, more studios, a pergola converted to a corridor, and a library. In 1981 the School of Journalism took over the building. "As soon as we moved in," they reported, "we felt cohesion".

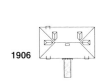

1906

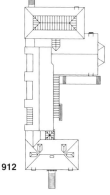

1908

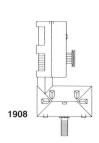

1912

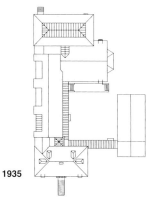

1935

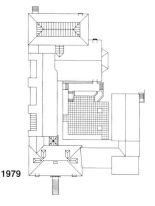

1979

of which deliberately isolate themselves from the real sources of feedback on building performance—lawyers and developers. Architects don't approve of lawyers or developers. They also diligently ignore facilities managers, the people most steeped in building function and malfunction. Frank Duffy told a conference of facilities managers, "I as an architect find facilities management liberating because it introduces the dimensions of time and use into buildings."[27] Author Fred Stitt observes, "For years architects and engineers have largely overlooked an enormous potential source of design work…. Facilities managers can give you long-term renovation and maintenance contracts on their buildings. Those contracts make you the 'house design firm' and lead directly back to major new building projects in the future."[28] Such a long-term relationship is exactly the sort that William Seale lauded in 19th century French architects.

In 1990 the American Institute of Architects polled its members on what competencies they would most like to develop. Out of the dozens offered, the next-to-last was "Develop facilities management services."[29]

Schools of architecture, in my limited experience, are wonderful and terrible. Wonderful because they foster the last great broad Renaissance-feeling profession, terrible because they do it so narrowly. They focus obsessively on visual skills such as rendering, models, plans, and photography. Sight substitutes for insight. The artistic emphasis discourages real intellectual inquiry, diverting instead into vapid stylistic analysis, and it shuts out the experience of non-artists such as engineers and real estate professionals.

Even riper for revolution than the schools are the architectural magazines. Any one of them could build circulation with a several-year crusade against the scandals in their industry, taking the perspective of the aggrieved users of buildings. People cannot believe that something so obviously important and permanent as buildings can be designed so badly. Unchallenged practices persist for decades—sliding glass doors and floor-to-ceiling glass partitions that people smash their noses on; extensive south- and west-facing windows that become solar ovens; extensive ground-floor windows that people invariably curtain for privacy; women's rooms designed symmetrically to be the same size as men's rooms, when double the size and facilities are needed; and the innumerable sad, dry, leaf-filled courtyard fountains that looked so cheery on the renderings but were too much trouble to maintain.

Architecture magazines could be the monthly avenue of feedback for the profession on what works and doesn't work. Instead they are the monthly barrier. What architects *are* kept well abreast of are the things which have advertisers—new materials and new

THE MEN'S ROOM of the Royal Institute of British Architects in London offered an unimpeded view of the urinals from the corridor when anyone went through the doors. A strategically-placed coat rack and modest panel corrected the error without bothering to conceal it. It's an endearingly inexpensive fix, no doubt instructive to young architects.

September 1990. Brand.

technology. That's helpful in a world of fast-moving technology, but what is even more needed than simple description is analysis—of reliability, life-cycle behavior, environmental impact, user acceptance, compatibility with other materials, ease of *dis*assembly—the sort of thing that *Consumer Reports* does for ordinary people.[30]

It would be nice to see architecture magazines invent an entertainingly journalistic form of post-occupancy evaluation of famous buildings new and old. Let careers rise and fall on those criteria for a change. People learn quickest from other people's mistakes, especially if the errors are assessed with compassion and humor. Prestigious design awards could be redirected. The American Institute of Architects' "25-Year Award" is on the right track because it honors time, but it pays no attention to use, only to abiding stylistic influence. Much better is the Rudi Bruner Award, which rewards social effectiveness and requires the jury to visit the candidate buildings. We need to honor buildings that are loved rather than merely admired. Admiration is from a distance and brief, while love is up close and cumulative. New buildings should be judged not just for what they are, but for what they are capable of becoming. Old buildings should get credit for how they played their options.

The needed conversion is from architecture based on image to architecture based on process. Some architects see it coming. Herman Hertzberger, in Holland, writes, "The point…is to arrive at an architecture that, when the users decide to put it to different uses than those originally envisaged by the architect, does not get upset and consequently lose its identity…. Architecture should offer an incentive to its users to influence it wherever possible, not merely to reinforce its identity, but more especially to enhance and affirm the identity of its users."[31]

Sir Richard Rogers affirms, "One of the things which we are searching for is a form of architecture which, unlike classical architecture, is not perfect and finite upon completion…. We are looking for an architecture rather like some music and poetry which can actually be changed by the users, an architecture of improvisation."[32]

Frank Duffy hectors his profession: "The reason I hate these architectural fleshpots so much is because they represent an aesthetic of timelessness, which is sterile. If you think about what a building actually does as it is used through time—how it matures, how it takes the knocks, how it develops, and you realize that beauty resides in that process—then you have a different kind of architecture. What would an aesthetic based on the inevitability of transience actually look like?"

The conversion will be difficult because it is fundamental. The transition from image architecture to process architecture is a leap from the certainties of controllable things in space to the self-organizing complexities of an endlessly raveling and unraveling skein of relationships over time. Buildings have lives of their own.

---

27 Francis Duffy, "Measuring Building Performance," *Facilities* (May 1990), p. 17.

28 Fred Stitt, *Designing Buildings That Work* (New York: McGraw-Hill, 1985), p. 107

29 The very least preferred was, "Apply knowledge of seismic construction." The next-to-last *pursued* competency was, "Perform post-occupancy evaluations." Joseph Bilello and Cynthia Woodward, "Results of AIA Learning Survey," *Architecture* (Sept. 1992), p. 100. The magazine had no comment on these findings.

30 An impressive example of this kind of reporting is *Environmental Building News* (bimonthly, $60/year from: EBN, RR1 Box 161, Brattleboro VT 05301).

31 Herman Hertzberger, *Lessons for Students in Architecture* (Rotterdam: Utgeverij 010, 1991), p. 148.

32 Sir Richard Rogers, "The Artist and the Scientist," in *Bridging the Gap* (New York: Van Nostrand Reinhold, 1991), p. 146.

Old House Journal (November 1991), p. 80.

1991 - Surreal estate in Queens, New York. A former duplex becomes a monoplex, and the owner of the left side shows he can do exactly what he damn well pleases with his property, while the owner a millimeter away to the right shows that he will not be budged. What the street lost in civility it gained in entertainment value, and how fortunate that no bland design review board got in the way. Architect Monty Mitchell took the photo for the *Old House Journal's* popular "Remuddling" feature.

CHAPTER 6

Unreal Estate

"PEOPLE WANT to get rich quick. That's at the bottom of every problem I've seen around buildings," says a carpenter friend, drawing on years of repair and remodeling work.

Part of it plays out in the all-too-brief design and building phases. Architect Peter Calthorpe affirms, "Rushing is at the root of all lack of quality." But haste, waste, and avarice distort buildings throughout their lives. Even if architects and builders were perfect, most buildings still would maladaptively freeze up or lose their way because of other pressures. These influences control the building from afar, not to evil ends but to distant ends, locally deadening. The building tenant lives in diffuse dread of the

**PROPERTY LINES CUT LIKE A RAZOR.
They are geometical, two-dimensional,
not of this world. But nothing else so
rigorously sculpts everything we build.**

remote landlord, the remote prospective buyer, the remote
building inspector and tax assessor.

Every building leads three contradictory lives—as habitat, as
property, and as component of the surrounding community. The
most immediate conflict is financial. Is your house primarily a
home or primarily an asset? Economists dating back to Aristotle
make a distinction between "use value" and "market value."[1] If
you maximize use value, your home will steadily become more
idiosyncratic and highly adapted over the years. Maximizing
market value means becoming episodically more standard, stylish,
and inspectable in order to meet the imagined desires of a
potential buyer. Seeking to be anybody's house it becomes
nobody's. Whole neighborhoods tend toward uniformity as
everyone avoids being the money-wasting best house on the
block or the sore-thumb worst house on the block.

A building is the interface between two human organizations—the
intense group within and the larger, slower, more powerful
community outside. The building's Site, Structure, Skin, and the
connection to its Services all are shaped by the community at
large. Agencies such as the planning board, design review board,
developer's office, and homeowners' association decree the size
and shape of your lot, where you can build or expand on it, the
size and shape of your building, the look of your building, and
what you can use it for. The overall rule is always "Fit in." It is
never "Become interesting." Some buildings become interesting

anyway. What you see on the street is the product of the
unending conflict between the organizations inside and outside—
buildings pretending to fit in or defying fitting in. Each facade
asserts, "Don't worry; nothing special happening in here!" or "I
found a way to be unique and there's nothing you can do about
it." I recall a student house in Berkeley with a neon sign in the
window flashing all night: "Smash the state."

One area of perpetual discord is the enforcement of building
codes. The earliest cities had them. In the Nineveh of 3000 B.C.
the Assyrian king Sennacherib decreed, "If ever any person living
in the city pulls down his old house and builds a new one, and the
foundation of that house encroaches on the royal processional
way, they shall hang that man upon a stake over his own house."[2]
As a youth I regarded building codes as the embodiment of all that
was unoriginal and constricting in society. Later I learned their
value.

In 1989 I happened to be in the middle of an earthquake in San
Francisco and wound up trying to help rescue a couple trapped in
a collapsed, burning building in the Marina District. The husband
managed to get out, the wife burned to death. Their four-story
apartment building had a non-code "soft story" of garages at the
street level which collapsed, and the old plaster-and-lathe walls
turned into kindling. As a California engineer put it, "Earthquakes
don't kill people. Buildings do." Thousands would have been
killed instead of a total sixty-two dead in that Richter 7.1
earthquake if the San Francisco Bay Area had not been

---

[1] "Aristotle made a very important distinction between 'oikonomia' and 'chrematistics.'
The former, of course, is the root from which our word 'economics' derives.
Chrematistics…can be defined as the branch of political economy relating to the
manipulation of property and wealth so as to maximize short-term monetary exchange
value to the owner. Oikonomia, by contrast, is the management of the household so as to
increase its use value to all members of the household over the long run." Herman E. Daly
and John B. Cobb, Jr., *For the Common Good* (Boston: Beacon, 1989), p. 138.

[2] H. W. F. Saggs, *Civilization Before Greece and Rome* (New Haven: Yale, 1989), p. 120.

aggressively researching and enforcing earthquake building codes for decades. (Lower stories must have plywood shear walls and be bolted to foundations, for example.) A 1988 Richter 6.7 earthquake in Argentina killed 25,000.[3]

It was fire codes that led the way in regulating the materials and design of buildings. After London burned in 1077, 1087, 1135, and 1161, an assize of 1212 laid down the law: "Whosoever wishes to build, let him take care, as he loveth himself and his goods, that he roof not with reed, nor rush, straw nor stubble, but with tile only, or shingle or boards, or, if may not be, with lead or plastered straw."[4] Buildings can't learn if they don't last. Most building code systems are a manifestation of the whole community learning. What they embody is good sense, acquired the hard way from generations of recurrent problems. Form follows failure.

Building codes are an adaptive and local phenomenon. There are 44,000 code-enforcement bodies in the US. These have been semi-standardized to three major regional systems—one each for the Northeast, the Gulf states, and the Midwest/West—and reflect to some degree historic and climatic differences. The codes often force builders and dwellers to act against their short-term interests, requiring heavy-duty wiring in case more electricity is needed later, strong walls in case extra stories are added later, and energy conservation that will pay off in lower utility bills over the long run.

This is an old and interesting problem in organizational learning. How do people learn to do cheap problem-prevention instead of expensive problem-cure?[5] There's no immediate reward for putting in a sprinkler system, only extra nuisance and expense. A larger, slower entity—the community—has to do the learning and instill the lesson, by convention, habit, rite, or law.

Convention is preferable to law, being more adaptive, accommodating, and locally appropriate, but a fast-moving society outruns the pace of informal convention and must resort to abstract law. Gene Logsdon points out the trade-off:

When building codes came to our unusually stable community, it was the more honest builders who fought them. They argued quite rightly that codes established minimal standards, and minimal standards sanctify mediocrity. Mediocre builders can undersell good builders. The upshot is that newcomers hire mediocre builders thinking themselves protected by codes, and they get mediocre houses. Natives continue to hire good builders and save money in the long run.[6]

At their worst, code enforcers block creativity and defy reason, answerable to remote abstractions that have nothing to do with the present case or opportunity. On the widespread estate lands of Chatsworth were a great many solid old stone farm buildings that Deborah Devonshire thought might be nice shelters—"stone tents"—for the teeming hikers and campers in the Peak District. She took the idea to the local authorities:

It seemed we might as well have been trying to do in a lot of innocent people. It appeared that anyone using the stone tents would most likely die from having no lavatories, no windows, no fire escapes, no heating, no water, no wash hand-basin for both sexes and no dustman calling every day. And could we assure the Authorities, in duplicate, that the floors would be level, that there would be a damp-proof membrane and the roof would be insulated and "the thermal requirements of the walls would be determined in individual buildings," whatever that may mean?… And then I read that "grouping together of persons in an unventilated building presents a grave risk of respiratory infection and it is totally alien to good housing." This made me feel very tired and old, and I retired from the scene in case I should get cross.[7]

After two years of effort by Chatsworth people, the ruling committees declared that the stone buildings were "movable dwellings" and were usable by Girl Guides as an exempted body. Eventually the idea was copied all over the Peak District, without restriction.

One of the major features of the ongoing life of any building is the hide-and-seek between remodelers, both amateur and professional, and the enforcers of codes—the building inspectors. Jamie Wolf, a professional remodeler in Connecticut, estimates with dismay that only 10-to-15 percent of all remodeling is ever formally given a permit and inspected. The techniques for hiding new work are legion: put stain on the stacks of new lumber and plywood so they are inconspicuous; instantly age the wood siding on new construction by rolling on a slurry of a few handfuls of ready-mix concrete in a few gallons of water; do just a little of the work at a time to avoid complaint or inspection; inside, put down old carpeting on new stairs; make new roof beams look ancient by staining them and blowing a shop vacuum full of dust and dirt on the wet stain. Most of it comes down to neighbor relations. Inspection evaders have the work done when the neighbors aren't around, and they make deals: if you don't complain about my new dormer, I won't say anything about the deck you snuck in last year.

The result is a mixed bag. Certainly more constant adjustment to buildings comes from all this informality, but every carpenter I've talked to complains that the thing they most dread in any remodeling job is shoddy prior work done on the sly and not up to code—"cob work done by some *handyman*." They contract to fix a warped bathroom floor and find they have to completely redo the plumbing, wiring, walls, and floor joists because earlier slapdash work put in hazardous wiring and leaky, rot-producing pipes.

People avoid getting permits for good reason. It may well trigger a tax reassessment. The permissions gauntlet can seem endless, especially if one's building is found to be in violation of recent (sometimes contradictory) ordinances. In New York City the cost of getting permission for remodeling often exceeds the cost of the work itself. Communities that want their built environment to improve over time would do well not to punish remodeling work. They could keep tax reassessment separate from improvement—do it strictly by calendar or only at the time of resale. They could revamp the permit and inspection process to become one of welcome guidance that helps people reduce costs and hassle. The procedure could even be linked to such services as tool-lending from the town library.

The community exercises its greatest control over a building by having decreed in advance how the property lines may go. The all-constraining Site dominates everything else about a building. City lots have been long rectangles with their short side facing the street since medieval times. Their size is enormously influential. Small lots, Anne Vernez Moudon found in her landmark study of San Francisco neighborhoods, make for constant fine-grain adaptation instead of the sudden, devastating changes that can come with large parcels.[8] "Everything depends on the pattern of ownership of the land," she writes. Small lots give greater individual control and thus greater variety, and they encourage more pedestrian activity. The more owners, the more gradual and adaptive the ongoing change. It's a conservative, wholesome kind of change—the place looks a little different every year, but the overall feel is the same from century to century.

3 Similarly, hurricane Andrew (August 1992) caused most of its $25 billion of damage to buildings that were constructed way below code, while code-compliant buildings mostly survived fine. Most of the damaged or destroyed buildings had no architect or engineer involved in their design, and the shoddy construction was never inspected properly. James S. Russell, "Tragedy of Andrew Rolls On," *Architectural Record* (August 1993), pp. 30-31.

4 Margaret Wood, *The English Mediaeval House* (London: Bracken, 1965), p. 292.

5 It's dental hygiene. Your teeth will fall out eventually if you don't brush and floss, but there's no immediate reward for doing that. You can get people to brush, if you convince them that it will make their smile look and smell nicer. But they won't floss.

6 Gene Logsdon, *The Low Maintenance House* (Emmaus, PA; Rodale, 1987), p. 11. See Recommended Bibliography.

7 Deborah Devonshire, *The Estate* (London: Macmillan, 1990), p. 129.

8 Anne Vernez Moudon, *Built for Change* (Cambridge: MIT, 1986), p. 188. See Recommended Bibliography.

1917 - In 1886 the Kann family opened a clothing store in the Second-Empire-style building (1875) on the corner of Pennsylvania Avenue and 8th Street. Two brothers named Saks then were expanding their clothing business in the new (1884) building on the right corner—it later became the famed Saks Fifth Avenue in New York. By 1917 Kann's had grown to occupy the whole block, punching through interior brick walls to join four buildings. Keep your eye on the building with cupola to the right.

1977 - Kann's prospered through constant innovation. It was the first Washington department store to offer highly discounted prices, full return on goods sold, credit to low-income customers, and it was the first to use black models in its ads. But with the recession of 1974 and pressure from city planners, Kann's had to close in 1975. It was boarded up and some of the aluminum facade was removed by 1977.

The community also rules a building by how it handles services—water, electricity, gas, sewerage, phone, television, and, peripherally, transportation. That these are each managed by different bureaucracies means that streets are torn up for service repair and replacement far too often. Some new towns in Sweden have tilt-up sidewalks with all the services laid out handily underneath; snow conveniently melts off them. "Neo-traditional" town planners are rediscovering alleys as a place to hide service lines, garbage pick-up, and cars (garage in back instead of dominating the front). Alleys also are kid country and a place for informal neighbor contact.

As for transportation, Joel Garreau's *Edge City* has the key insight: "Cities are always created around whatever the state-of-the-art transportation device is at the time. If the state of the art is sandal leather and donkeys, you get Jerusalem…. The combination of the present is the automobile, the jet plane, and the computer. The result is Edge City."[9] Unplanned new urban centers at junctions of the auto corridors around the periphery of older cities are a phenomenon which has spread rapidly around the industrialized world. The most auto-defined building of all is the cheap roadside office building. Built to a formula of 250 square feet inside per employee and 400 square feet outside per employee's car, it is always a one-story building covering 40 percent of the lot, with bleak asphalt parking on the remaining 60 percent.[10]

9 Joel Garreau, *Edge City* (New York: Doubleday, 1991), p. 32. See Recommended Bibliography.

10 Title above, p. 120.

11 Kostoff noted that Rome's whole city plan wound up focusing on Il Duce's personal window in Palazzo Venezia. Mussolini removed thousands of homes to visually isolate Rome's ancient monuments. Hitler enjoyed his planning role so much he had a secret passage to the office of his 31-year-old master architect, Albert Speer, later Armaments Minister. Had their grand avenue plans come to full fruition in Berlin, 80,000 homes would have been destroyed

1979 - A mysterious fire on February 2, 1979, ended the debate over whether to save the old buildings.

1981 - For most of a decade the space remained empty, awaiting redevelopment. The old Fireman's Insurance Company building at the far right still had no cupola.

CONSOLIDATING LOTS alters urban change from steady and small to sudden and large—from adaptive to convulsive. Kann's Department Store on Pennsylvania Avenue in Washington, DC, grew from its corner building on the left, gradually taking over the adjoining buildings until it absorbed the Saks building on the right corner. A unifying aluminum facade went up in 1961. Under duress from subway construction and restrictions by the federally-mandated Pennsylvania Avenue Development Corporation, Kann's closed in 1975. Preservationists tried to save the old commercial buildings, but a 1979 fire attributed to arson forced demolition. By 1990 a monumental new complex called Market Square was completed.

1991 - Market Square, a huge multi-use public-private complex, is an agreeable place, designed by Hartman-Cox. The open space and part of the building has a US Naval Memorial with fountains and sculpture. In the building (and in a symmetrical one facing it across 8th Street) are restaurants, shops, prestige offices, and, on the top four floors, two hundred condominium apartments. In the spirit of preservation, a cupola was restored to the building on the right.

Edge Cities and one-story strip buildings are the bane of city planners, who often view themselves as architects writ large. City planners came into existence in America after 1893 and have been accumulating power ever since. The late architectural historian

Spiro Kostoff used to lecture his students that by the far the most ambitious and successful city planners in the 20th century were Benito Mussolini, who transformed Rome, and Adolf Hitler, who set about transforming Berlin.[11]

The planning profession keeps oscillating between being destructively radical and destructively conservative. The "urban renewal" disasters of the 1950s and 1960s led to eloquent repudiation by Jane Jacobs's epochal *The Death and Life of Great American Cities* (1961), but their influence lives on in the bloom of homeless people on city streets in the 1980s—there were no "slum" hotels left. So the new suburban developments went in the opposite direction. Garreau defines master planning there as "that attribute of a development in which so many rigid controls are put in place, to defeat every imaginable future problem, that any possibility of life, spontaneity, or flexible response to unanticipated events is eliminated."[12]

American planners always take inspiration from Europe's great cities and such urban wonders as the Piazza San Marco in Venice, but they study the look, never the process. Garreau asked social historian Dennis Romano and planning historian Larry Gerkins about Venice. Romano: "Those who now romanticize Venice collapse a thousand years of history. Venice is a monument to a dynamic process, not to great urban planning.... The architectural

harmony of the Piazza San Marco was an accident. It was built over centuries by people who were constantly worried about whether they had enough money."[13] Gerkins: "The Piazza San Marco was not planned by anyone.... Each doge made an addition that respected the one that came before. That is the essence of good urban design—respect for what came before."[14]

You can see respect for what came before in the much-lauded public-housing approach taken by Charleston, South Carolina, starting in 1983. The housing authority distributed 113 units on infill sites (vacant lots) all over the city, with the new buildings looking exactly like the old, many of them replicating the famed

CONTINUITY OF USE is a common real-estate effect. Long-term property owners such as religious groups consider the site essential, the buildings dispensable. The Presbyterians in Santa Fe, New Mexico, have tried three churches so far at their corner on Grant and Griffin Streets.

ca. 1977 - The leading practitioner of "Santa Fe style" was architect John Gaw Meem (see pp. 141-150). In 1939, following a city program to make Santa Fe look more Spanish colonial, he reclad the Presbyterian bricks to look like the very Catholic Spanish mission church (1815) at Ranchos de Taos. In that building the heavy buttresses support bell towers; here they're decorative.

ca. 1881 - In 1867, Presbyterians in Santa Fe took over a crumbling adobe Baptist church and renovated it with Gothic windows and a crenelated bell tower.

ca. 1913 - Rebuilt of brick in 1912, the church Gothicked its windows yet further.

Ben Wittick. Museum of New Mexico. Neg. no. 15855.

J. Weltmer. Museum of New Mexico. Neg. no. 15175.

Arthur Taylor. Museum of New Mexico. Neg. no. 111923. These photos appear in Sheila Morand, *Santa Fe Then and Now* (Santa Fe: Sunstone, 1984), pp. 84-85.

vernacular Charleston single house. The subsidized tenants took proud care of their homes and blended amiably into their new neighborhoods.

The success of the Charleston housing planners flies in the face of the basic injunction of city planners since the 1910s: keep functions and classes separate. Zoning was introduced in Berkeley in 1914 and New York City in 1916 to protect residential neighborhoods from commerce, industry, and undesirable immigrants. Soon, legal barriers clattered into place all over the cities—residential here and here, commercial center in there, industrial out yonder. No more unsanitary high-density living areas (there *had* been a health problem). No more mixed living and working (there had been nightmarish sweatshops). And no melting pot; these cities would be stratified strictly by economic class, creed, and race. It would all be efficient and just, because experts were in charge.

Zoning must have worked to some degree, since it has lasted so long. But it succeeded in freezing up cities so tight that new Edge Cities out on the periphery became inevitable, and something was lost with all that comfortable stasis. Luxemburger planner-apostate Leon Krier declares, "Functional Zoning is not an innocent instrument; it has been the most effective means in destroying the infinitely complex social and physical fabric of pre-industrial urban communities, of urban democracy and culture."[15] People find ways around zoning ordinances—quietly setting up home businesses in their garage or basement, quietly moving into industrial lofts—but like barrio dwellers they can succeed only so far before authority discovers and curtails them. Quelling change, zoning quells life.

I'm lucky enough to live in a neighborhood where zoning has broken down, partly because it's on the border between city and county jurisdictions. A couple of decades back, what had been zoned as a light industrial waterfront area in Sausalito, California, was invaded by illegal residential houseboats such as mine, eventually over 400 of them (now mostly legal). The result is mildly scruffy and utterly convenient and neighborly. From my door it is a short walk not only to my office but to: public storage, auto repair, boat supplies, bike supplies, office supplies, film supplies, tire service, a car-battery manufacturer, a gas station, a gym, a notary public, a supermarket, a convenience store, a deli, and seven restaurants—all in the low-rent, Low Road, malleable part of town. I visit friends in nice homes elsewhere and it feels as if they live in a desert, zoned out of a walkable way of life, stuck in a place where nothing ever changes.

The progress from legal to illegal usage is worth study, because it shows how communities learn from their buildings. A few years back I lived inexpensively in a tiny cottage in an extremely affluent place called Belvedere, California. Two local women were pressuring the city council to register and tax all of the town's "second units" (also called accessory apartments, granny flats, mother-in-law units) and outlaw all new ones. Most of the innumerable second units had been added quietly over the years without permits or official notice by the city. The council was about to pass the new ordinance when front page coverage in a local paper brought the biggest crowd in memory to a city council meeting. The two women sat rigid as the whole town explained in detail that second units were the salvation of Belvedere. They gave families the flexibility to stay in their homes, because there was a place for the aging parent, the au pair, the growing teenager. They provided affordable housing for city staff, local nurses, local shop employees—the town's whole support population. Rent from second units reduced the cost of primary homes. A financially and socially brittle community was made broader and more adaptive by second units. Don't outlaw them, help them.

[12] Joel Garreau, *Edge City*, p. 453.

[13] Title above, p. 10.

[14] Title above, p. 217.

[15] Leon Krier, "Houses, Palaces, Cities," *A.D. Profile 54* (London: Architectural Design, 1984).

The Belvedere council unanimously reversed its ruling. According to the Urban Land Institute, about 40 percent of US communities have amended their zoning to permit second units, and a few promote them.[16] Parts of Europe encourage second units. Neo-traditional town planners are even incorporating second units into new construction, sometimes as a flat over the garage on the alley.

But most new communities seek to pre-empt any such adaptivity by repressive, fiercely enforced "covenants, conditions, and restrictions." These are the dread "CC & Rs" that homeowners' associations use to control such details as what colors you may paint your house, what pets (and in some cases what children) you may keep, how your lawn will look, your roof, your fence, your driveway (no campers, trucks, or car repair), your backyard (no drying laundry or unstacked firewood).[17] Any neighbor might report you. What if you ignore or defy such rulings? The home-owners' association can take your house or send you to jail. Joel Garreau points out that these organizations have all the powers of government—the ability to tax, to legislate, and to police—without the usual restrictions of democratic representation or being answerable to the US Constitution.[18]

The homeowners' associations keep growing in numbers and power. In 1990 there were 130,000 of them in America. In California some 70 percent of new developments have home-owners' associations running them. They're obviously desired, but are they desirable? Garreau contrasts a new development such as Irvine, California, to the once-deplored original Levittowns that were created for postwar families back in 1949:

> The old Levittowns are now interesting to look at; people have made additions to their houses and planted their grounds with variety and imagination. Unlike these older subdivisions, Irvine has deed restrictions that forbid people from custom-izing their places with so much as a skylight…. Owners of expensive homes in Irvine commonly volunteer stories of not realizing they had pulled into the driveway of the wrong house until their garage-door opener failed to work.[19]

What makes homeowners' associations so viciously conservative? Market value is determined not by how well a house works, but how it looks in the context of its neighborhood—"curb appeal," as it's called. Vast effort has gone into making the development look nice to a carefully calculated market segment, and that must not be undermined. When you sell your nice house (Americans move every eight years, on average), do you want the prospective buyer to see someone repairing their car or putting out laundry to dry next door? Suppose they've got a metal roof instead of tile, or a nonstandard dormer sticking out? Well, if they can't, you can't. This degree of institutionalization of real estate value over use value is odious enough as an invasion of privacy, but it also prevents buildings from exercising their unique talent for getting better with time.

Real estate is an astonishingly unexamined phenomenon. Books on the history of architecture outnumber books on the history of real estate 1,000 to 0, yet real estate has vastly more influence on the shape and fate of buildings than architectural theories or aesthetics. In America, according to the *Wall Street Journal*, "The businesses of financing, building, selling and furnishing real estate account for nearly one-fifth of the nation's total output."[20] This one-fifth, unfortunately for national stability and building longevity, is the last arena of truly wild-eyed financial speculation. Real-estate bubbles inflate and pop, inflate and pop with amnesiac regularity. Since the boom-times are as destructive as the busts, you'd think that governments and banks would take steps to gentle the oscillation. Instead they feed it.

In both America and Britain a normal real-estate boom was in progress by the beginning of the 1980s. Britain's was fueled in part by its notorious "long lease," which locks a tenant in for 25 years and requires that rents may only be revised upwards. Very attractive investment, that. Prime Minister Thatcher's government worthily privatized public housing (the council houses), and suddenly there was a large and naive population of homeowners to whom mortgages looked like free money.

Meanwhile, in the States in 1981, new Reagan-era tax laws generously subsidized real-estate investors with such goodies as accelerated depreciation, and property became the leading tax shelter. At the same time, the Savings & Loans banks (thrifts) were deregulated but federally insured, so they lost all discipline and poured money into real-estate schemes. Also the pension funds and insurance companies decided that real estate was the place for their institutional money. None of this had anything to do with property fundamentals—actual demand, location, building type, access, quality of design. Property went up because property went up. Decades' worth of new buildings were being built overnight because money was available, not because tenants were waiting.

A huge secondary market in mortgages built up, completely abstracted from any familiarity with actual properties or owners. Predictably, the weakest loan paper migrated to the weakest or least expert institutions. Then came the needle from the government, America's 1986 Tax Reform Act, which sensibly but suddenly took away all of the 1981 real-estate tax loopholes. New commercial construction practically ceased. Cities were dotted with "see-through" high-rise office buildings with no one in them. Property prices sagged, then tumbled. A whole generation of mortgages and land and construction loans went sour and took the Savings & Loans with them, which brought in the all-insuring government, which didn't have the money either. (In Britain the lending banks foreclosed and then had no idea what to do with the empty buildings). While the governments fibrillated, the national economies tipped into severe, prolonged recession. New construction could not perform its traditional rescue because too much future had been oversold and overbuilt. All we had built was a bigger bubble than ever.

Even the up part of the real-estate cycle is crippling to buildings. People get into a "trade up" mentality about their houses and treat them as investments. Any improvements made are for the imaginary next buyers, not themselves. The homeowners' associations or zoning bodies savagely resist any degradation of value brought on by such aberrations as, say, a smaller, cheaper house on the block. When all the houses are investments, nobody will waste money being nicer than the rest of neighborhood (since neighborhood determines value), and nobody will be permitted to be less nice, so they're stuck in lockstep, commodities on the shelf.

Downtown it's worse. The financial pressure on every square foot is so great that the value of land overwhelms the value of existing buildings, and they become disposable. Highrises, which multiply land value with swift efficiency, darken the skies. The extreme case was Tokyo in the 1980s, when its real estate market spiraled

---

[16] "Granny Flats: Boon to Rental Market?" *Land Use Digest* (July 1990), p. 4.

17 . The CC & Rs of Irvine, California, state that they are "for the purpose of uniformly enhancing and protecting the value, attractiveness and desirability of the Properties." An excerpt gives the flavor:

"Section 7.04. Parking and Vehicular Restrictions. None of the following (collectively 'Prohibited Vehicles') shall be parked, stored or kept on any street (public or private) within the Residential Area: any commercial type vehicle (including, but not limited to, any dump truck, cement mixer truck, oil or gas truck or delivery truck); any recreational vehicle (including, but not limited to, any camper unit, house/car or motor home); any bus, trailer, trailer coach, camp trailer, boat, aircraft or mobile home; any vehicle not in operating condition or any other similar vehicle; any vehicle with a width in excess of eighty-four (84) inches; any trash dumpster; or any vehicle or equipment, mobile or otherwise, deemed to be a nuisance by the Board. No Prohibited Vehicle shall be parked, stored or kept on any Lot or Common Area except wholly within an enclosed garage, and then only if the garage door is capable of being fully closed. Prohibited Vehicles shall not be allowed in any driveway or other exposed parking areas, or any street (public or private) within the Residential Areas, except for the purpose of loading, unloading, making deliveries, or emergency repairs.... *Garages or other parking areas within the Residential Area shall be used only for parking authorized vehicles, and shall not be used for storage, living, recreational, business or other purposes.*"

I added those italics. I had to.

[18] Joel Garreau, *Edge City*, p. 187.

19 Title above, p. 271. The cultural historian Paul Groth says the critics were dead wrong about the Levittowns: "They've survived beautifully. People are proud about adapting them. The original cheap materials wore out, as predicted, and people were happy to put in new materials."

The garage-door experience has been turned into a tool. You drive down your street of identical houses with the garage door opener pressed on. The house with its garage door opening is yours.

[20] Paulette Thomas, *Wall Street Journal* (17 May 1991), p. 1.

so high that the total value of Japanese real estate was worth four times the total value of America's—a nation twenty-five times its size. Highrises were being torn down in downtown Tokyo after five years, because the land value was so high it was trivial to demolish the last over-specified skyscraper and put in the next over-specified one. The average life of a building in Tokyo was seventeen years.

This is why cities devour buildings. Commercial centers act like gravity wells, with everything nearby getting sucked into their vortex. The closer a building is to the center, the more endangered. Nothing is exempt. Architecture critic Ada Louise Huxtable observed:

> There is no art as impermanent as architecture. All that solid brick and stone mean nothing. Concrete is as evanescent as air. The monuments of our civilization stand, usually, on negotiable real estate; their value goes down as land value goes up.[21]

The form of cities persists for centuries, but their intense economic metabolism consumes their physical substance. Frank Duffy: "In a British city we add to the stock of space about 2 to 3 percent a year, which means you renew a city every fifty years. Of course North American ones are much shorter time-scale." Leon Krier: "Sixty percent of German buildings survived the Second World War. Only less than 15 percent of these survived the industrial plans of the last thirty years."[22]

Cities are just as destructive when the real estate market dives. Lower value means less rent, hence less maintenance, which leads

---

[21] Ada Louise Huxtable, "Anatomy of a Failure," *Will They Ever Finish Bruckner Boulevard?* (New York: Times Books, 1970), quoted in *All About Old Buildings* (Washington: Preservation Press, 1985), p. 15. See Recommended Bibliography.

[22] Leon Krier, "Houses, Palaces, Cities," *A.D. Profile 54* (London: Architectural Design, 1984), p. 102.

Society for the Preservation of New England Antiquities. Neg. no. 15796-B. The first two photos appear in Peter Vanderwarker, *Boston Then and Now* (New York: Dover, 1982), pp. 46-47. See Recommended Bibliography.

ca. 1880 - Haberdashery dominated Winter Street in the late 19th century. With shops at the street level and residences above, the 1820s buildings of brick and granite had acquired some 1870s embellishments such as the handsome oriel (bay window) on the corner building.

CONTINUITY OF USE is apparent even in high-turnover commercial districts, where buildings can't keep up with intense traffic and changing land values. At the corner of Tremont and Winter Streets in Boston, facing the subway stop on the Boston Common, only two buildings out of seven survived from 1880 to 1980, and only one out of ten businesses from 1980 to 1993.

1980 - Only the two corniced brick buildings on the left were still around by 1980. Jewelry, furs, shoes, and fast food for tourists and shoppers had joined the mix. An 1887 building with heavy stonework dominated the corner.

1993 - When I rephotographed the street, most of the businesses had changed since 1980, but many were doing the same thing as their predecessors. Fanny Farmer Candies was still on the corner. Above it, Edward's Fur Shop had turned into Frank's Custom Tailor, and Loring Studios had become a coin and stamp shop. Next door, Hayat Shoes was now Jewelry Plus, and Dee's Jewelry became Au Bon Pain (toney fast food). Chock Full O' Nuts turned into Boston Grill & Bar-B-Que and then Rami's Kosher Hot Dogs. Miles Postcards and Paperbacks was taken over by Talal Shoes. Homer's Jewelry, after being sold and gutted, reopened as the Diamond and Jewelry Center, and a Payless Shoe Source went into the formerly empty shop next door. City planners had converted Winter Street to a mall-like pedestrian thoroughfare by repaving with bricks, discouraging cars, and encouraging street vendors.

to even lower rent. If the trend goes far enough, eventually it's not worth it to the owner to have any tenant at all, and the building burns in an insurance fire or it stands empty, begging to be demolished. Some neighborhoods sink beyond even demolition, their properties worthless and abandoned to gangs. Zero value land destroys buildings as surely as infinite value.

As usual, the rate of change is everything. Plummeting real-estate value is devastating, and soaring real estate value paralyzes homes and guts commercial districts. But in a slow down market, people stay where they are, improving their property for themselves and becoming real denizens of their neighborhoods. In a slow up

1976 - COMMUNITY-ROOTED buildings thrive. The local branch of the Bank of America had been in Point Reyes, California, since the 1920s when the branch manager, Tom Molinari, decided it was time to expand out of their cramped old building with its poor heating and minimal parking. Across the street was a vacant corner lot, formerly a gas station. Molinari was the kind of banker who made house calls to customers and regularly schmoozed with the community leaders. Here a number of them are, lined up for the ground-breaking at the new site—contractor Joe Redmund, local Coast Guard commander Dick Manning (many of the CG wives would work in the bank), grand old man of the area Toby Giacomini, manager Molinari, Don DeWolfe from the Lion's Club, and Ed Vaca, president of the local businessman's association.

market, they are rewarded for rehabilitating marginal structures, gradually turning factory lofts into artists' studios or townhouses, retaining a civilized urban mix.

The perennial driver of haste is what is called "the time value of money." To be an attractive investment, a building must earn at a rate greater than the interest paid on its purchase and construction loans. The project has to make money fast. More developers are done in by the payments on their loans (the "carry," as it's known and feared) than by any other cause. Thus they insist on blitzkrieg construction schedules, and then they shovel in the tenants to get some rent flowing. The result: instant, shallow, flashy buildings with no adaptivity and no investment for the long term.

Everything about money and buildings says they are expected to live only a brief time, typically 30 years. Most mortgages go for just 25 or 30 years, and so the asset life comes to match the finance period. Just when you own it, you have to replace it. US tax laws require a residential building to be depreciated over 27.5

years, a commercial building over 31.5 years. The value of the building has supposedly arrived at zero after three decades, and you can write that mythical "cost" off your income during the 27.5 years but not afterward. Small wonder buildings are constructed to last only that long.

Loans and depreciation discount the future. Thus forests are cut and turned into money, because you can make more money on the money than you can on the forests logged sustainably. *The Economist* declares, "The very concept of discounting, when applied to the environment, is based on the idea that all resources belong to those alive today."[23] It is a hasty, shabby way of living that results. Buildings that survive and adapt can teach the opposite lesson—*the use value of time.* Jane Jacobs sings a Poe-like hymn to the value of older, somewhat decrepit buildings in cities:

> Time makes the high building costs of one generation the bargains of a following generation. Time pays off original

Bank of America

1988 - Still around for a rephoto twelve years later were Manning, Giacomini, Molinari, and Vaca. Under pressure from the Design Review Board, the building was styled to look "like a saloon" to fit in with older buildings on the street. Molinari carried the theme into the interior with unbankish lace curtains, flocked wallpaper, captain's chairs, and local historic photos. In the 1990s, Bank of America pulled out of its branches in small towns, and the nearby Bank of Petaluma took over the building and most of the accounts and staff that went with the building. The photos by Art Rogers are part of his famous series called "Yesterday and Today," which chronicles changes and continuity in the lives of people in and around Point Reyes.

capital costs, and this depreciation can be reflected in the yields required from a building. Time makes certain structures obsolete for some enterprises, and they become available to others. Time can make the space efficiencies of one generation the space luxuries of another generation. One century's building commonplace is another century's useful aberration.[24]

Old buildings *are* more freeing.

Suppose we heeded the lesson. We would adjust building economics to reflect and serve long-term value—higher use value,

finer-grained market value. The flow of money through a building acts to organize that building. Will the building be organized around the moment of sale or the decades of use? Jane Jacobs draws the distinction between "cataclysmic money and gradual money," noting that cataclysmic money in cities is destructive whereas gradual money is wholesome and adaptive. Chris Alexander agrees: "The money is wrong in most buildings, and it's crucial. There should be more in basic structure, less in finish, more in maintenance and adaptation. Once a building heads downhill, you lose motivation to fix it. You have to maintain a steady flow of money into a building, and mortgages skim that."

Mortgages are an interestingly mixed issue. And a huge issue. In California alone, mortgage debt in 1990 added up to $437 billion, equal to 65 percent of California's gross product.[25] What's good about mortgages is that they make ownership easier, and ownership serves long-term responsibility and maintenance. American and British tax laws lavishly reward mortgages (you can

[23] "The Price of Green," *The Economist* (9 May 1992), p. 87.

[24] Jane Jacobs, *The Death and Life of Great American Cities* (New York: Random House, 1961, 1993), p. 247. See Recommended Bibliography.

[25] "California dreaming, on a rainy day," *The Economist* (23 June 1990), p. 77.

deduct the interest from your income), and quasi-government institutions with the mock-friendly names Fannie Mae and Freddy Mac seek to stabilize the mortgage market. As a result, an impressive two-thirds of America's 94 million households own their own homes.[26] The stabilizing value of mortgages is even clearer overseas. In Japan and Germany—nations highly regarded and highly rewarded for taking the long view—mortgages are commonly written to amortize over 100 years. The financing spans three generations. Japan further discourages commoditization of houses with tax laws that punish rapid purchase and sale.[27]

So far, so good. But Chris Alexander argues that a mortgage-bought building tends to be an over-packaged illusion of completeness that defeats any kind of incremental approach.[28] Two out of every three dollars spent on the purchase of the building go into paying interest—a horrific drain even with the tax advantages. A $200,000 home winds up costing $600,000 by the time it's all paid for. During real-estate boom years, people will invest even more than the usual 25 percent of their income in house financing, betting as high as 40 percent. It is the opposite of sound investing, as a professional home improver points out: "It takes years and years to build up any significant equity through payments alone. The best way to build equity quickly is to buy wisely and remodel to add value."[29]

Buildings do better over time when they are closely held and closely cared for. Most are not. Quick-buck speculation and

abstract investment is the norm. You look at a typically chaotic city street and you see the artifacts of remote institutions blind to each other, even secretive. Finding out who actually owns commercial properties is often nearly impossible. The buildings usually have passed through many owners, each one concerned primarily with the interests of backers and potential buyers, seldom the building users, the neighbors, or the public in the street. The absentee landlord is a traditional villain for good reason. In Britain particularly, most developers sell a new building as quickly as possible to an institution such as a pension fund or insurance firm—distant, inept, and adversarial landlords. Tenants are locked into the 25-year "long lease" with full responsibility for maintenance, which they shirk.

Landlord and tenant are automatically in conflict, especially with houses and apartments. The landlord wants stable tenants who will treat the property as their own, *but* will make no alterations or improvements without the landlord's approval. The tenant wants to spruce up or adapt the place, *but* gains nothing in equity thereby; in fact the rent may go up. Every detail of repair, maintenance, or improvement becomes grounds for an argument. The standoff is so well understood to be deleterious that banks usually refuse to give loans on homes that are not owned by the occupants.

When the landlord is the state, as it was in communist lands, you get the ultimate in negative maintenance. All visitors to the mortally rundown buildings of Eastern European nations have tales like Brian Eno's: "My wife and I were checking in to a hotel in Moscow. Our host showed us to our room, and began switching on the lights. As he turned on the one by the door, a great tongue of flame issued forth from a light fitting in the ceiling. He calmly switched it off again and said, 'Don't use that one.'" Since no one owned the light, why should anyone fix it? A command economy displaces responsibility even further outside the building than a market economy does.

A final real-estate effect on buildings is the stress caused by

---

[26] U.S. Bureau of the Census, in *Housing and Market Statistics*, National Association of Home Builders (May 1992), p. 44.

[27] The real estate market in Japan is made less volatile by taxing a seller *150 percent* of the capital gain if a building is sold within two years, and 100 percent within five years. *The Economist* (8 Dec. 1990), p. 9. Sweden has similar tax regulations.

[28] Alexander's argument against mortgage financing appears in detail in Stephen Grabow, *Christopher Alexander* (Boston: Oriel, 1983), pp. 144-5.

[29] Lawrence Dworin, *Profits in Buying and Renovating Homes* (Carlsbad, CA: Craftsman, 1990), p. 63. See Recommended Bibliography.

LEON KRIER:

Leon Krier, "Houses, Palaces, Cities," *A.D. Profile 54* (London: Architectural Design, 1984), p. 36.

Two Forms of Accumulation

...and only one has history in it.

constant turnover. Homes are occupied by new owners every six-to-eight years on average. Most apartments keep tenants for only three years at a time. A set of offices is likely to suffer ten or more tenant organizations in its lifetime. Each turnover usually is accompanied by a complete renovation or remodeling. Often there are two renovations per ownership turnover—the first by the departing owners looking to jazz up the place for a higher sale price. But that one is wasted, because the new owners immediately make another extensive renovation to match their own tastes and needs. The building can't learn much with all those shock treatments. On the other hand, turnovers can help upgrade some of the basics such as roof, foundation, and services. The increasing use by buyers of private building inspectors in recent years has helped this process.

A downward sequence may not be all bad, either. Each turnover might help degrade the building toward Low Road status, where the landlord can no longer be bothered to care what the tenant is doing with the place, and sufficiently busy inhabitants such as artists or small businesses may begin a turn toward gradual rehabilitation of the building and its neighborhood. Adaptive life re-enters the building when it becomes too cheap for speculation.

Turnover refreshes, but it also erases. Nearly everything about real estate estranges buildings from their users and interrupts any form of sustained continuity. A triumph of abstraction, real estate operates distant from the daily life of building use, distant from the real. The "real" in "real estate" derives from *re-al*—"royal"—rather than from *res*—"thing"—which is the root of "reality." Realty is in many ways the opposite of reality.

All that is solid melts into cash. Real estate turns buildings into money, into fungible units devoid of history and therefore of learning. A room would gradually embody its history, but a dollar gains its time-and-space binding power from having no history, and the dollar wins. The room becomes not what happened in it, but what its square-footage is worth. Monetization frees from history by destroying history. Almost nothing can stand against this universal liquidity. But two ways have been found. One is a community response that deserves its own chapter (coming right up). The other is just a wiser approach to investment.

"People want to get rich quick." The other side of the coin is, *Go broke quick*. Real estate is the classic case of soar and collapse, of tycoons going bankrupt and taking shortsighted banks with them. Work done in haste is necessarily shoddy, a house of cards. On a go-fast schedule there is no margin for a single error, and error is inevitable. High risk, high loss.

The opposite strategy is much surer, because the errors are piecemeal and correctable. When you proceed deliberately, mistakes don't cascade, they instruct. Low risk *plus time* equals high gain. This strategy treats the fundamentals of the living investment with attention and respect. The lesson of realty laced with reality is: "Get rich slow."

These photos courtesy of the Planning Department, Town of North Hempstead, NY, and the McDonald's Corporation.

1991 - Thanks to preservationists, the McDonald's just off the Jericho Turnpike on Long Island, New York, is in a classy old house. Is that great, or pathetic? I think it's fine.

CHAPTER 7

Preservation: A Quiet, Populist, Conservative, Victorious Revolution

AMERICA'S foremost architectural historian, Vincent Scully of Yale, called it "the only mass popular movement to affect critically the course of architecture in our century."[1] He was speaking of the historic preservation movement, which swept seemingly out of nowhere in the 1970s and 1980s to reverse everything that had been done to the built environment in the 1950s and 1960s. Modernist architecture, urban renewal, go-go real estate—all were suddenly treated as the enemies of civilization and beaten back. People *liked* old buildings, and professionals who couldn't get along with that could go find another line of work.

How did such a profound change come about? Why wasn't it noticed in the media? How has it changed the way buildings are treated?

Preservation was one of the swiftest, most complete cultural revolutions ever, yet because it happened everywhere at once, without controversy or charismatic leadership, it never got the headlines of its sibling, the environmental movement. Also its reward cycle was much quicker, and therefore quieter, than environmentalists could count on. Retro *worked*; preservation

1926 - Around 1860, Augustus Denton remodeled the old (1795) vernacular house on the left as the kitchen, dining room, and servants' quarters for a grand new Georgian center-hall house with eight big rooms and handsome porches. Augustus later was a town supervisor, bank director, and treasurer of the Jericho Plank Road that became the turnpike that eventually attracted McDonald's.

ca. 1985 - After 1926, the Denton house was turned into a restaurant. As usual, one restaurant led to another, including a "Charred Oak" and a "Dallas Ribs." By 1985, neglect and dilapidation had made the house a candidate for landfill. Augustus's gable-end chimneys, porch railings, and semicircular-headed attic windows were long gone.

1991 - McDonald's paid a million dollars for the site and then learned they had to restore the building. Back came the chimneys, the railings, and the attic windows. Local preservationists let them add a single-story addition hidden behind the house, so the place could seat 140. It was the 12,000th McDonald's to be opened.

FAST FOOD IN A SLOW HOUSE. Begun in 1795, brought to Georgian magnificence in 1860, the Denton House was slated for demolition in 1987 when aroused local residents got the building designated as historic and protected. A deal was made with the new owner, the McDonald's chain, to restore the house to the condition visible in a 1926 photo. Now a high-tone burger joint, it honors both the original building and the lineage of restaurants that have occupied the place since 1928.

Large houses are exceptionally skilled at being comfortable, being loved, and being adaptable. Like old factories and warehouses, they are always prime candidates for preservation's best political-economic-design device—"adaptive use."

paid off, and the movement could demand ever more, based on proven success. James Marston Fitch, who founded the first historic preservation training program at Columbia University in 1964, could assert in 1990, "Preservation is now seen as being in the forefront of urban regeneration, often accomplishing what the urban-renewal programs of twenty and thirty years ago so dismally failed to do. It has grown from the activity of a few upper-class antiquarians…to a broad mass movement engaged in battles to preserve 'Main Street,' urban districts, and indeed whole towns."[2]

A typical success story of the period was *Old House Journal.* In 1967 Clem Labine's wife, Claire, talked him into buying an 1883 brownstone rowhouse in Brooklyn, New York. Labine remembers, "I finally rationalized the purchase on the idea that

[1] Vincent Scully, Charles Moore Gold Medal Presentation, 6 Feb. 1991, at the National Buildings Museum, Washington, DC.

[2] Quoted in "Focus on Preservation," *Architectural Record* (March, 1991), p. 152. Fitch wrote the authoritative text *Historic Preservation*—see Recommended Bibliography.

since it's masonry, it should be low-maintenance, therefore I won't have to spend all my time working on the house. Not only did I spend all my time working on the house, I eventually built a career telling other people how they can spend all their time working on their house." The Labines discovered a network of young couples fixing up old houses, and on their kitchen table in 1973 they started a 12-page newsletter. At that time the word "Victorian" was a synonym for bad taste. Within two years *Old House Journal* was making money. By 1992 it had a circulation of 150,000 old-house fanatics, many of them retired people as well as the young preservers.[3] Improving real-estate value was a surprising by-product, never the main idea. Clem Labine: "I've always approached this field from the angle of cultural responsibility. To me the two most important operative words in dealing with buildings are respect and sensitivity."

A building nostalgia industry took off, purveying everything from claw-foot bathtubs to stamped-metal ceilings. Labine remembers, "In 1976, when I published the first census of the historical products industry, there were only 205 companies whose products could be considered historically styled. By 1991 there were nearly 3,000 historical products suppliers, and the list was growing too fast to keep up with." Architectural salvage of such things as antique plumbing, old decorative windows, and handsome paneling was earning $100 million a year,[4] and that was just the legitimate companies. Midnight salvage in empty old buildings became the newest form of vandalism, ironically the antithesis of respect and sensitivity.

When I began research for this book I was drawn immediately to the preservationists, because they are the only building professionals with a *pragmatic* interest in the long-term effects of

time on buildings. They work creatively with the economics and changing uses of buildings, and they promote expertise in the crafts of longevity. Architectural historians, on the other hand, had almost nothing for me. As a subset of art historians, they are interested only in the history of intention and influence of buildings, never in their use. Like architects, they are pained by what happens later to buildings. Building historians are the opposite. They are the ingenious detectives who deduce how specific buildings have changed through the years, but still there is no design or operational payoff unless they hook up with preservationists.

Preservationists have a philosophy of time and responsibility that includes the future. They are passionately interested in the question, "What makes some buildings come to be loved?" and they act on what they learn. The result is a coherent, still-evolving ethical and aesthetic body of ideas. One architect has observed, "Preservation has become the best carrier of that moral force architecture needs if it is to have value beyond shelter. Preservation is capable of projecting a vision of new possibilities, of hope for our own future, which functionalist modern once claimed for itself and which has now fled from that style."[5]

There is more to continuity of the physical environment than just habit and nostalgia. Old buildings embody history. They are worlds; in old buildings we glimpse the world of previous generations. The cultural historian Ivan Illich remarked once, "History gives us distance from the present, as if it were the future of the past. In the spirit of contemplation it releases us from the prison of the present to examine the axioms of our time." Old buildings give us that experience directly, not through words.

What does this have to do with aesthetics? "Beauty is in what time does," says Frank Duffy. Something strange happens when a building ages past a human generation or two. *Any* building older than 100 years will be considered beautiful, no matter what. Having outlived its period of being out of fashion, plus several passing fashions since that, it is beyond fashion.   If it has kept

---

[3] The bimonthly *Old House Journal* costs $24/year from 2 Main Street, Gloucester, MA, 01930. England has a quite separate but also excellent magazine with the same name.

[4] Sally G. Oldham, "The Business of Preservation is Bullish and Diverse," *Preservation Forum*, National Trust for Historic Preservation (Winter, 1990), p. 16.

[5] Robert Jensen, "Design Directions: Other Voices," AIA Journal, May 1978, quoted in *All About Old Buildings* (Washington: Preservation Press, 1985), p. 30.

1884 - Oklahoma pioneer Frederick Severs founded the first general store in Okmulgee in 1868. In 1882 he rebuilt the place with sandstone and expanded it to two stories.

BACKLASH. The 1950s and 1960s changed the look of old buildings to new. The 1980s changed them right back. Downtowns successfully countered the suburban shopping mall boom by turning themselves into historic theme parks. Through its "Main Street" program, the National Trust for Historic Preservation was able to employ government incentives such as tax credits to get downtown merchants to band together and out-mall the malls. Typical of hundreds of projects, the Severs Block in Okmulgee, Oklahoma, was renovated backward to 1906.

ca. 1920 - Severs prospered, and by 1906 the old general store had been incorporated in the two-story brick facade of Severs Block—the north side of the 100 block of East 6th Street.

1985 - In 1954 the Citizen's National Bank, which had occupied the building since 1906, modernized and unified their corner structure with stucco, marble panels, and roman brick. Even the windows were radically changed, which is rare.

High Road continuity, the whole place is highly adapted, complex and mysterious, a keeper of secrets. Since few buildings live so long, it has earned the stature of rarity and the respect we give longevity.

Durability counts for more and more as our decades grow hastier. "One of the things that people like about older buildings," says Clem Labine, "is that they were built to last. They have a sense of permanence about them. Up until the 20th century, people thought that they were building for the ages." Traditional materials like brick, stone, stucco, slate, and wood show age

1991 - In 1988, motivated by Okmulgee's Main Street project and at considerable expense (sweetened by a 20 percent tax credit), the bank tore off the stucco and marble and roman brick and scrupulously restored the original brick and sandstone to match the 1920 photograph. Even the windows were changed back.

attractively, whereas recent materials such as aluminum, plastic, and exposed concrete age ugly. Often they don't hold up as well as the older materials—innovative adhesives lose their grip,

FANTASY VERSUS REALITY VERSUS PRESERVATION. British preservation is renowned (and often resented) for its attention to extreme detail in discouraging alterations to "listed" buildings. The strictures are a barrier to adaptivity in buildings, and they feel to the inhabitants like an invasion of their privacy and property rights. On the other hand, the stringency reflects a century of experience with preservation. Seemingly small incremental changes to old buildings can accumulate over decades with drastic results. These drawings and captions are from Pamela Cunnington's excellent *Care for Old Houses* (see Recommended Bibliography). She suggests that additions never be larger than half the size of an original house, if its charm and cohesiveness are worth preserving.

hidden metal fasteners corrode, and the veneer and curtain walls of even recent Modernist buildings fall off. Surveying the rustic stone buildings of the Chatsworth estate, the "clerk of works" told the Duchess, "Nothing as good will ever be built again, so let us make proper use of what we've got."[6]

Widespread revulsion with the buildings of the last few decades has been an engine of the preservation movement worldwide. Shoddy, ephemeral, crass, over-specialized, the recent buildings display a global look especially unwelcome in tradition-enriched environs. Auto designer Robert Cumberford in France: "The Dordogne is still astonishingly beautiful, but it is in places disfigured by the 20th century. It took a few years to realize that *everything* I found ugly was put in place since 1910. We've got a lot to answer for with our cheap, nasty constructs." Planner and architect Leon Krier in Italy: "Perhaps we would love this country all the more if the architecture constructed after 1930 were to vanish; but the majority of architects refuse to think about the profound significance of this fact."[7] Preservationist Paul Goldberger: "A lot of our belief in preservation comes from our fear of what will replace buildings that are not preserved; all too often we

DEATH OF A COTTAGE

"We found this lovely little thatched cottage—just what we were looking for."

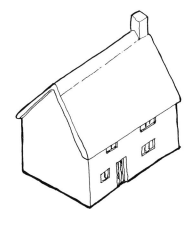

"Of course, there was no bathroom so we had to build one, and we added a utility room at the same time. The council made us put large windows in the new part."

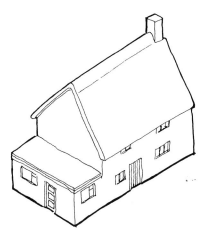

"We found the cottage was really too small, so we built on a new kitchen and extra bedroom at the back."

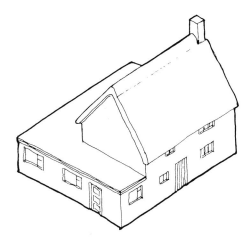

fight to save not because what we want to save is so good but because we know that what will replace it will be no better."[8]

But you can get only so far with cultural and aesthetic arguments. The only effective game in town is real estate. At first reluctantly, later with relish, preservationists learned to think and act like developers and property owners in order to recast economic incentives in favor of preservation. The crunch for every building comes at the time of the rehabilitate-or-demolish decision, brought on by real-estate pressure or building obsolescence—usually both. Much of the time the decision is a close call that could go either way.

[6] Deborah Devonshire, *The Estate* (London: Macmillan, 1990), p. 110.

[7] Leon Krier, "Houses, Palaces, Cities," *A.D. Profile 54* (London: Architectural Design, 1984), p. 10.

[8] Paul Goldberger, *Preservation: Toward an Ethic in the 1980s*, quoted in *Landmark Yellow Pages* (Washington: Preservation Press, 1990), p. 95.

[9] According to Donald Rypkema, a real-estate economist who specializes in preservation issues, even extensive rehabilitation (services, windows, roof) typically costs 3 to 16 percent less than demolishing and replacing an old building. "Making Renovation Feasible," *Architectural Record* (Jan. 1992), p. 27.

"Old buildings *save* you money," preservationists tell the investors, developers, and city councils. Often you can buy an old building cheap. Rehabilitation of an old building may be expensive, but it's still significantly less than comparable new construction.[9] You can avoid the expense, disruption, and environmental burden of demolition. The rehab work often can proceed by stages, while part of the building is still profitably occupied, and it can take less time than new construction. It uses less materials, so you're protected from rising materials costs. And most older buildings are energy-efficient enough to be brought up to within 80 percent of state-of-the-art efficiency with inexpensive work to the windows and roof.

If the costs of rehabilitating an old building still exceed its expectable market value—a circumstance known as "the gap"—the public often is willing to step in and sweeten the numbers with tax credits, tax abatements, low-cost loans, historic districts, transferable air rights or development rights, and other incentives. The government (local, state, federal) is persuaded to do this by economic as well as cultural arguments. Existing public services, such as utilities and public transportation serving the old building,

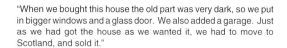

"With the children growing up we needed more bedrooms, so we put another storey on the back addition. Then, when the children married and left home, the house was too big for us and we sold it."

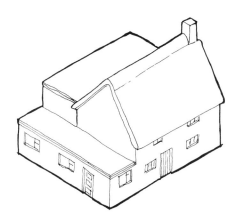

"When we bought this house the old part was very dark, so we put in bigger windows and a glass door. We also added a garage. Just as we had got the house as we wanted it, we had to move to Scotland, and sold it."

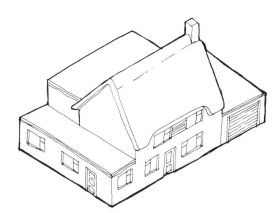

"When we bought this house the thatch was in a very bad state, so we decided to have it tiled. We needed more rooms, so we built up the front at both ends. We put a new Georgian-style door and windows in the front, to be more in keeping with an old house."

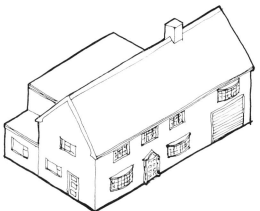

are utilized without new expenditure. The enormous energy expense "embodied" in the building is saved.[10] Rehabbed buildings can help revive a whole neighborhood, even a whole town or city, with resulting higher rents and tax revenues. The least attractive part of town may become suddenly the most attractive. A revitalized neighborhood attracts new investment, new business, and often the pure gold of tourists.

The bond between old buildings and tourists is absolute and venerable. Think of any famous city in the world and you view a mental slide show of the characteristic look of the buildings vernacular to that place from an earlier time. Tourists have helped revive or save many a building or neighborhood that was ready to be discarded by locals. Of course no one respects tourist opinion (though they should). What they respect is tourist money.

A British real estate magazine reported, "Britain's architectural heritage 'is almost certainly England's most valuable tourism asset,' according to Save Britain's Heritage. In 1988 domestic visitors generated £7,850 million [$12 billion]."[11] In America, each of the 1,000 house museums in the country with a budget over $50,000 was estimated to generate $6 million of economic activity locally—a total of $6 billion nationwide, not even counting the additional 4,000 smaller house museums.[12] The aggregate numbers are overwhelming: *The Economist* reported in a special report on tourism and travel that it is the world's largest industry— $2 *trillion* in sales annually, employing 6.3 percent of the global workforce. Two-thirds of that number is estimated to be "travel for pleasure."[13] And it keeps growing; world tourism is expected to double by 2010 over what it was in 1993.[14]

Preservationists might be thought of as tourists-in-place. They bear a pilgrim's veneration for the local ancient buildings, whether "ancient" is measured in decades or millennia. They will lie down in front of bulldozers to save a 1930s Art Deco bus station. In view of this kind of devotion and the whole complex of behavior and organization that has grown to support it, cultural historian Chris Wilson regards preservation as a "secular religion." How did

such a phenomenon arise? Is it likely to last?

The tradition that America draws on took shape among the romantics of early 19th century France and England. The French archaeologist A. N. Didron stated in 1839 the slogan that still guides all preservationists: "It is better to preserve than to repair, better to repair than to restore, better to restore than to reconstruct."[15] Later the French architect Eugene Viollet-le-Duc, while elaborately restoring medieval works such as the cathedral of Notre Dame, simultaneously was devising the theoretical foundations of functionalist Modernism—an affiliation of the passionate and the pragmatic worth reviving.

In England two varieties of romantic came into open warfare. Gothic Revival enthusiasts armed with Victorian wealth and confidence set about "restoring" everything dark and stony to an Early (13th century) Gothic look regardless of the actual age and tradition of the building. John Ruskin, the romantic's romantic, an aesthete as revered in America as he was in England, rebelled. In *The Seven Lamps of Architecture* (1848) he thundered that restoration "means the most total destruction which a building can suffer: a destruction out of which no remnants can be gathered: a destruction accompanied with false description of the thing destroyed."[16] Inspired by Ruskin, the artist and activist William Morris founded in 1877 the Society for the Protection of Ancient Buildings (SPAB), which inveighed against the "restoration tragedy" and founded the preservation movement in England. (Another scion from the same root was the Arts and Crafts movement, which medievalized new construction and artifacts with a sophisticated simplicity, craftsmanship, and warmth still imitated.)

In the intensely public debate it was Victorian "scrape" versus Ruskin's and Morris's "anti-scrape"—tear off the plaster to expose ancient stones (even if they were plastered originally) versus Leave The Building Be, including the original plaster and everything that was added later to keep the building working. "Scrape" won the 19th century, "anti-scrape" the 20th.

Preservation is now a national pastime in England. The National Trust, founded in 1894, is the largest private landowner in the country, with 1 percent of its total land and 10 percent of its coast. It also administers 200 country house estates, most of which came to it via Draconian inheritance taxes introduced after World War II. English Heritage, a government conglomeration, takes care of 350 properties, including Stonehenge, and also administers the selection and protection of "listed" historic buildings—6,000 Grade I ("exceptional"), 23,000 Grade II* ("particularly important"), and 400,000 Grade II ("special interest"). In addition, there are scores of volunteer national and local organizations ranging from the Vernacular Architecture Group to the Friends of Friendless Churches. William Morris's SPAB has spun off one sub-organization specializing in Georgian buildings and another defending Victorian structures—including some Victorian "improvements" to ancient buildings that Morris originally founded SPAB to fight against.

The cumulative effect of such widespread preservation activities—amateur, professional, and governmental—is a country that feels solidly rooted in its own history, culture, and place. The entire south of England, Chris Alexander opines, is one of humanity's most beautiful artifacts, intricate and refined, historically deep, an aesthetic whole. Both tourists and natives can explore the landscape, not confined to the occasional cathedral or Tower of London. Everywhere they prowl are buildings that still work for a living, richly textured, expert at being exactly where they are and what they are, visibly cherished.

America's preservation movement began in patriotism rather than romanticism. Around 1850, George Washington's Mount Vernon was offered by the family for $200,000 to the state or federal government, neither of which could imagine paying that much for a decaying house and farmed-out plantation. In 1853 a 37-year-old southern gentlewoman, medically frail and single, heard of the impasse and wrote a shaming letter, "To the Ladies of the South," published in the Charleston *Mercury* and reprinted nationwide. Thus Ann Pamela Cunningham inspired and soon deftly organized

the Mount Vernon Ladies Association with female fundraisers in every state. They purchased Mount Vernon in 1858, and they run it to this day. A pattern had been set. In matters of preservation, organizations of volunteers would take the lead.

A series of patriotic places followed—President Andrew Jackson's home, a famed Meeting House in Boston, Paul Revere's house, colonial Williamsburg. In 1910 a Society for the Preservation of New England Antiquities was set up, patterned on William Morris's SPAB in England. In 1931 Charleston, South Carolina, came up with an innovation to head off a rash of gas stations replacing beloved old buildings. They declared most of downtown an "Old and Historic District" with special protections for all its buildings. Now there are hundreds of such districts.

Then, starting in 1947, the proliferating preservation groups recognized that they needed unified clout. The National Trust for Historic Preservation was established in 1949 with an official charter from Congress. At first the organization collected grand country homes in the manner of Britain's National Trust, but in 1966 a new dimension was opened with the publication of a single book, a masterpiece of pre-legislation propaganda titled *With Heritage So Rich*. In a book replete with photographs of

---

[10] "…The amount of energy initially invested in a building—equivalent to about 12 gallons of gasoline per square foot—is enough to heat, cool, and light the same building for more than 15 years…. Construction of new buildings in the United States accounts for more than 5 percent of the total US energy use each year." William I. Whidden, "The Concept of Embodied Energy," *New Energy from Old Buildings* (Washington: Preservation Press, 1981), p. 130.

[11] Richard Catt, "A Few Guidelines to Putting a Price on Architectural History," *CSW [Chartered Surveyor Weekly]* (1 Aug. 1991), p. 18.

[12] Sandra Wilcoxon, "Historic House Museums: Impacting Local Economies," *Preservation Forum*, National Trust for Historic Preservation (May 1991), p. 10.

[13] Special section on travel and tourism, *The Economist* (23 Mar. 1991), p. 5.

[14] According to the World Tourism Organization. Edward Epstein, "World Insider," *San Francisco Chronicle* (14 May 1993), p. A10.

[15] Always cited to *Bulletin Archeologique*, vol. 1, 1839.

[16] John Ruskin, *The Seven Lamps of Architecture* (New York: Dover, 1848, 1880, 1989), p. 194.

splendid old buildings—some of them being demolished on camera—was a full account of preservation progress and problems, laced with inspirational prose such as Walter Muir Whitehill's:

> In describing the temple of Hera at Olympia in the second century A.D., the traveler Pausanias noted that the style of the temple was Doric and pillars ran all around it, and that in the back chamber one of the two pillars was of oak. Originally the columns of this ancient building had been of wood, but in the course of seven centuries preceding Pausanias's visit, all save this one had been replaced with cut stone. Presumably this single oak column had been allowed to remain out of piety, as a visible symbol of the antiquity of the temple on this site.[17]

Most of the findings and recommendations in the book quickly found their way into law—the National Historic Preservation Act of 1966. At last the federal government joined the volunteers: the National Park Service went into partnership with the National Trust for Historic Preservation. A national registry of historic buildings was set up. A system of state preservation organizations was empowered. Tax benefits would aid approved rehabilitation. Federal money would flow. Suddenly America, with very little history to preserve, had a preservation policy and apparatus as effective as any in the world. The newspapers barely noticed.

Most of the support for the National Trust still came from private contributions, endowments, and the dues of its 250,000 members. In 1992, $28.5 million in private money was matched by $5.7 million in grants from the government. Preservation in America has a broad base of support, but much of the leadership has come from the wealthy. They're the ones with the time, taste, influence, money, generosity, and concern for generations of time. Old money likes old things, and new money imitates old money. More than with other forms of philanthropy, the donors get to join in themselves—save a public building and help enjoy it, save a private building and live in it. In Britain Deborah Devonshire was bemused: "Who could have foreseen that the most fashionable dwelling houses of today would be those once occupied by animals or their food? …Stables, coach-houses, granaries, barns, kennels, dairies and sheepfolds are the smart addresses now."[18]

To offset the brute force of the real-estate market, the arbitrariness of city planners, and the trendiness of architects, preservationists had to come up with all manner of stratagems—easements, transferable development rights, tax increment financing, conservative design review boards, down zoning, and unholy coalitions of supporters who would never agree on anything but their affection for some shabby old building. In Hartford, Connecticut, the handsome Old State House (1796) was preserved by taxing every one of the 2,753 windows that looked upon it $5 a year. When old downtowns were emptied by retail traffic migrating to the new shopping centers on cheap land at the edge of town, the preservationists retaliated by instituting a "Main Streets" movement nationwide that taught downtown merchants how to imitate the organization and styling of the malls and lure the business back. With spectacular tax benefits they goaded developers into reinhabiting big old buildings and whole blocks with major new commercial centers, some of them celebrating historic themes.

It was the tax-credit gambit that did the most in sheer volume and that most impressed other countries. New federal tax laws in 1981 allowed developers a whopping 25 percent tax credit on the cost of rehabilitating certified historic buildings, and 20 percent on *any* building more than thirty years old, plus accelerated depreciation. Investment swarmed to such projects, and by 1987 some $14 billion had been spent on 21,000 historic buildings in 1,800 towns. It was too good to last. Included in the tax reforms of 1986 that sent the real-estate market plummeting was a reduction in the

---

[17] The book is back in print. Walter Muir Whitehill, "Promoted to Glory," *With Heritage So Rich* (Washington: Preservation Press, 1966, 1983), p. 137.

[18] Deborah Devonshire, *The Estate* (London: Macmillan, 1990), p. 109.

1871 - As the nation approached its first centennial in 1875, the search was on for a female hero of the Revolution. Philadelphia had two candidates—Lydia Darragh, an intrepid and effective spy, and Betsy Ross, who sewed American flags and *might* have sewed the first one. But Darragh's house on Second Street was replaced by a hotel, so that left Ross. A campaign to save her house hustled pennies from the nation's schoolchildren. Darragh was forgotten by history. Betsy Ross entered legend.

1987 - The house at 239 Arch Street became a national shrine in 1937. In 1975 the body of Elizabeth Ross Claypole, which had occupied two previous cemeteries without fanfare, was reverently interred next door.

FLAG HOUSE. **One benefit of America's late-19th-century enthusiasm for preserving buildings of patriotic significance is that some interestingly random buildings got saved. No other quality would have singled out the tiny three-story building on Philadelphia's Arch Street for veneration if it had not reputedly been the home of Betsy Ross, who reputedly received a commission from George Washington to sew the first United States flag. It now has half a million visitors a year.**

benefits to commercial rehabilitators that reduced their activity by two-thirds in the late 1980s.

One lasting effect of the boom years was that everyone in the building businesses became familiar with an intelligent set of guidelines which had to be met to get those tasty tax benefits. "The Secretary of the Interior's Standards for Rehabilitation" went through continuous stages of refinement as preservers and developers fought over their applicability. The 1992 version is worth printing here as a distillation of a century of anti-scrape wisdom and good advice for anyone interested in extending the life and value of a building in a High Road direction:

1) A property shall be used for its historic purpose or be placed in a new use that requires minimal change to the defining characteristics of the building and its site and environment.

2) The historic character of a property shall be retained and preserved. The removal of historic materials or alteration of features and spaces that characterize a property shall be avoided.

3) Each property shall be recognized as a physical record of its time, place, and use. Changes that create a false sense of historical development, such as adding conjectural features

ca. 1948 - The Count of Sully had a handsome mansion built in 1624 on the Rue Saint Antoine in the Marais district of Paris. Over the centuries, urban intensity and density filled in its stylish void and multiplied the floor levels inside.

ca. 1965 - As part of admirable government programs to preserve and rehabilitate the older parts of Paris, work began in 1951 to restore the Hôtel de Sully. Appropriately, it now houses the Caisse Nationale des Monuments Historiques. I wonder if its inviting gap will ever be filled in again some day.

SEVERE RESTORATION. If this Paris building were in America, Paragraph 4 of the "Secretary's Standards" would have encouraged keeping most of the elements added since 1624 instead of carving it back to its Baroque origins. I would agree. The building seems to have traded its soul for a false virginity out of keeping with its Marais setting.

or architectural elements from other buildings, shall not be undertaken.

4) Most properties change over time; those changes that have acquired historic significance in their own right shall be retained and preserved.[19]

5) Distinctive features, finishes, and construction techniques or examples of craftsmanship that characterize a historic property shall be preserved.

6) Deteriorated historic features shall be repaired rather than replaced. Where the severity of deterioration requires replacement of a distinctive feature, the new feature shall match the old in design, texture, and other visual qualities and, where possible, materials. Replacement of missing features shall be substantiated by documentary, physical, or pictorial evidence.

7) Chemical or physical treatments, such as sandblasting, that cause damage to historic materials shall not be used. The surface cleaning of structures, if appropriate, shall be undertaken using the gentlest means possible.

8) Significant archeological resources affected by a project shall be protected and preserved. If such resources must be disturbed, mitigation measures shall be undertaken.

9) New additions, exterior alterations, or related new construction shall not destroy historic materials that characterize the property. The new work shall be differentiated from the old and shall be compatible with the massing, size, scale, and architectural features to protect the historic integrity of the property and its environment.

10) New additions and adjacent or related new construction

[19] Many have noticed the contradiction in Paragraph 4, which declares pretty baldly, "Old change good, new change bad." This aesthetic revisionism will turn up again around vernacular buildings.

1865 - In Washington DC, the northwest corner of 19th Street and Pennsylvania Avenue, N.W. was already historic by 1865. In the row of federal townhouses known as "the Seven Buildings," President James Madison and Dolley occupied the corner building (built about 1795) during the years the White House was being rebuilt after the British burnt it in 1814. When this photograph was taken, Civil War General M. D. Hardin had his headquarters in the Seven Buildings. The occasion of the photograph, attributed to Matthew Brady, may have been President Lincoln's funeral.

ca. 1948 - Long used by government departments, after the Civil War the Seven Buildings became more commercial, with shops on the ground floor and offices and flats upstairs. By 1948 at least five of the buildings were still intact. The second from the right had added a story and the fourth acquired a mansard roof and three dormers while the sixth was converted in 1898 to a four-story apartment building. This series of photos is from the collection of Washington chronicler Charles Suddarth Kelly.

1981 - Only two of the original Seven Buildings survived the highrise boom of the 1970s, dwarfed between eleven-story and eight-story office buildings. Peoples Drug kept its corner location.

1989 - An infill office building replaced the 1898 apartment building and loomed over facade remnants of the two original buildings, both tidied up to look vaguely like the 1865 photo. Inside, it's all one building.

"FACADISM" is the dirty word preservationists use for projects that save the illusory fronts of old buildings to mask entirely new construction. The passerby doesn't know whether to be insulted by the crude lie or delighted by the surreal kitsch.

1990 - Heroic engineering was required to preserve the facades of the six-story Busch Building (1882, left) and two-story Kresge Building (1918, right) near Pennsylvania Avenue and 8th Street in Washington. The new mixed-use complex, which gutted the entire block, is eleven stories high, with five stories underground for parking, a health club, and a theater.

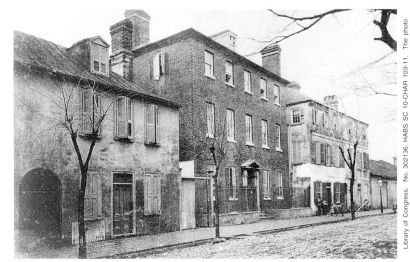

1883 - Wealthy rice planter Daniel Heyward had a brick "double house" (center building) constructed in 1772 on Charleston's lower Church Street. His son Thomas was a signer of the Declaration of Independence and leased the house to George Washington for a week in 1791. During the 19th century the building was used as a boarding house and a bakery.

1990 - The Heyward House was acquired by the Charleston Museum and, after a shop on the ground floor was erased, it was turned into the city's first house museum in 1930. The building to the right was a negro tenement known as "Cabbage Row" in the 1920s when the neighborhood's liveliness inspired Dubose Heyward to write the novel *Porgy* (1925). With the set of "Catfish Row" based on this building, the story became a play (1927), a George Gershwin opera (1935), and a movie (1959).

shall be undertaken in such a manner that if removed in the future, the essential form and integrity of the historic property and its environment would be unimpaired.[20]

(The Low Road, of course, has no use for such preciousness. Its route to authenticity is through directness, not time and continuity. The Improviser's Standards would read: "Mess with the building as needed until it works.")

What does preservation preserve? You might say it honors peculiarity, specific to the building and to the locality. It fights the invasive uniformity of franchise fast-food joints, multinational gas stations, and highrise office buildings. There is also the romantic attraction of participating in the local myth. City-watcher Kevin Lynch has noted:

> Patriotism and literary glamour have defined certain classic periods whose traces are most worth preserving: the late colonial and Revolutionary years in New England, the brief episode of pioneering in the forested interior, the ante-bellum days in the South, the period of exploration and cattle raising on the Great Plains (which passed so quickly), the

POVERTY STOPS CHANGE. The best-preserved downtown in America—Charleston, South Carolina—was saved by a combination of calamity and tradition. The calamity was the Civil War (1861-65), which impoverished the American South for decades. "Poverty is the best friend of preservation," says Clem Labine. "When property owners don't have a lot of money, they're no longer potential customers for the aluminum siding salesman or the fly-by-night contractor hawking the latest fad." In Charleston, which was full of the townhouses of plantation gentry, the tradition was to keep the home in the family no matter what—no improvements, few repairs, and hang on. When the city came back to economic life in the 20th century, it found itself with 3,600 suddenly priceless historic buildings. The buildings had survived that crucial second thirty years when a building is out of fashion, out of repair, and most vulnerable to demolition.

Detroit Publishing Company, Library of Congress. No. 314322, LC-D401-72484. Published in *Charleston: Come Hell or High Water* (Charleston: Whitelaw, 1976), p. 88.

1906 - A block south of the Heyward House on Church Street was Charleston's oldest house (left foreground), built about 1720 by Colonel Robert Brewton.

20 October 1990. Brand.

1990 - The look and human-scale feeling of Charleston is unchanged in a century. Realizing its value, in 1931 the city created America's first Historic District, now 1.25 square miles in size. Block after block in every direction looks just like this. It is a tourist bonanza, but the city has also kept its identity and original population, solidly rooted in its old buildings.

mining era in the Western mountains, the years of the Spanish colonies in the Southwest, and, of course, the undefined background of the scattered and 'timeless' Indian.[21]

While the regions provide myth, cities express history. The most spectacular example in this century of a city preserving its story by preserving its buildings is the Polish capital of Warsaw. World War II left the historic city a desert of ruins and rubble, with 88 percent of its buildings destroyed. But hidden away from the shelling in a provincial town was an exhaustive inventory of Warsaw's historic buildings—photographs and measured drawings that had been made in the 1930s. Over the next eight years Warsaw painstakingly reconstructed itself in its own past image as a vivid symbol and tool of the rebirth of Poland. James Marston Fitch observed, "The workmanship in *new* construction leaves much to be desired. Be that as it may, there is little evidence of poor workmanship in Polish *preservation* work. Here, the quality ranges from good to superlative."[22] So skilled had the Polish restorers become that they quickly got work all

over Eastern Europe reviving other demolished cities.

Comparing the divergent fates of bombed-out cities after the war, Fitch draws a general lesson:

> The radically remodeled centers of LeHavre, Rotterdam, and Dresden are dull, monotonous, auto-dominated cityscapes,

---

20 US Department of the Interior, *The Secretary of the Interior's Standards for Rehabilitation & Illustrated Guidelines for Rehabilitating Historic Buildings* (Washington: Government Printing Office, 1992), p. vi. See Recommended Bibliography. The National Park Service has a precise definition for *rehabilitation*: "the process of returning a property to a state of utility, through repair or alteration, which makes possible an efficient contemporary use while preserving those portions and features of the property which are significant to its historic, architectural, and cultural values."

21 Kevin Lynch, *What Time Is This Place?* (Cambridge: MIT, 1972), p. 30. See Recommended Bibliography.

Cultural historian Chris Wilson has a less romantic interpretation. He sees preservation as in part a bourgeois control scheme. Being against mixture and change, old white families promote purity and stability through shaping buildings to reflect the pre-industrial order of things. The South goes pre-Civil-War. New England goes pre-immigrant. California goes Spanish Colonial and carefully excludes contemporary Hispanics. Aware of the charge, the National Trust for Historic Preservation has developed a new emphasis on cultural diversity in its programs.

22 James Marston Fitch, *Historic Preservation* (Charlottesville: Univ. of Virginia, 1982, 1990), p. 382. See Recommended Bibliography.

with scarcely a visual hint to indicate their points of origin in space and time. In Nuremberg, Warsaw, and Leningrad the reverse is true: their visual specificity has been maintained for the pleasure of visitor and resident alike. Merely to *be* in their dazzling panoramas of Gothic, Renaissance, Baroque, Rococo, and Art Nouveau is an unforgettable experience.... In [preservationists'] practice, if not yet in their theoretical formulations, one can discover an incrementalist approach to urban development that closely parallels the theories of such iconoclastic planners as Jane Jacobs and Christopher Alexander.[23]

A similar contrast was played out in Fort Worth, Texas, in the 1970s and 1980s. One member of the oil-wealthy Bass family invested in blocks of new highrises. His competitive brother responded by putting money into preserving and reviving nearby blocks of old Fort Worth buildings. By the 1990s the highrises were partially empty and in trouble, while the preserved blocks had become the center of city life.

Likewise in Berlin. When the Wall came down, East Berlin was found to have a splendid collection of glorious old Prussian buildings amid the socialist junk. They were in horrible repair, of course. Author Pamela McCorduck noted, "Part of reunification is fixing these up again, especially since the West Berliners, in a frenzy of Modernism, tore down everything, even the buildings that might have been saved, and everything on that side looks like it appeared one night in 1955. They see the wonderful old buildings of East Berlin as part of their patrimony."

Maybe there's such a thing as an "aesthetic infrastructure," and it's as essential to city health as utilities, sewerage, and a transportation net. The present needs a past to grow on, according to Kevin Lynch: "Longevity and evanescence gain savor in each other's presence.... We prefer a world that can be modified progressively, against a background of valued remains, a world in which one can leave a personal mark alongside the marks of history."[24] Even the totally new Edge Cities out on the

POVERTY STOPS CHANGE. "Compton Wynates is said by many to be the most perfect Tudor house in England.... That [it] has survived so remarkably intact is perhaps due to a collapse of the family fortunes in the latter half of the 18th century. Spencer Compton, eighth earl of Northampton, came close to bringing about its total destruction. Having bankrupted himself through gambling and fighting a recklessly extravagant Parliamentary election in 1768, he fled to Switzerland, from where he issued orders for the demolition of Compton Wynates since he could no longer afford to keep it up. Luckily his faithful steward, John Birrell, who loved the house as if it were his own, chose to ignore these instructions. Instead he blocked up most of the windows to avoid the window tax, disposed of the contents by sale, and kept the place patched up for posterity as best he could. For the better part of a century the house lay empty, thus almost entirely escaping the major alterations that so often took place under the Georgians and Victorians. Poverty and neglect proved a blessing in disguise, preserving the atmosphere of absolute timelessness which still exists at Compton Wynates today." (Christopher Simon Sykes, *Ancient English Houses*, p. 107. See Recommended Bibliography.)

periphery may need this savor nearby in order to prosper. Joel Garreau is convinced that a relationship between the new periphery and the old core is essential to both: "Tourism is now the number one industry in New York City—ahead of financial services. It is also the fastest growing.... It is as if our downtowns have become antiques, in the best sense of that word. Edge Cities may represent the everyday furniture in our lives, but we recognize the downtowns as something to be cosseted and preserved."[25]

1808 - Around 1515, William Compton, First Gentleman of the Bedchamber to Henry VIII, was one of the first in England to build in brick since the Roman era. His lavish manor house in Warwickshire incorporated roof timbers and decorative windows from the demolition of Fulbroke Castle (15th century), which had been given to him by Henry VIII. By the time of this engraving, the original moat and outside courtyard had been removed, so the house no longer looked medieval.

ca. 1975 - Thanks to hibernating right through the 19th century, Compton Wynates was spared the customary Victorian additions to country houses—service wing, billiards wing, entrance hall, additional stories, clock tower, and overall restyling to early Gothic. The present Marquis of Northampton resides in the place, and it is closed to the public.

What alchemy turns a bad old building into a good old building? Vernacular building historian J. B. Jackson insists that a form of death must precede rebirth: "There has to be that interval of neglect, there has to be discontinuity; it is religiously and artistically essential. That is what I mean when I refer to the necessity for ruins: ruins provide the incentive for restoration, and for a return to origins."[26] But busy cities seldom tolerate ruins, and wood ruins can't survive the elements long. Jackson's version of reincarnation works best on masonry buildings in the countryside or in old factory and warehouse districts.

The most common form of survival of old buildings into renewed value comes from what preservationists have learned to honor and promote as "adaptive use." When a building designed for one purpose is put to a completely different use, its value deepens, says Jane Jacobs:

> Among the most admirable and enjoyable sights to be found along the sidewalks of big cities are the ingenious adaptations of old quarters to new uses. The town-house parlor that becomes a craftsman's showroom, the stable that becomes a

house, the basement that becomes an immigrants' club, the garage or brewery that becomes a theater, the beauty parlor that becomes the ground floor of a duplex, the warehouse that becomes a factory for Chinese food, the dancing school that becomes a pamphlet printer's, the cobbler's that becomes a church with lovingly painted windows—the stained glass of the poor—the butcher shop that becomes a restaurant.[27]

How many such violent transitions can a building endure? Any number, suggests Jacobs's account of a building in Louisville,

[23] Title above, p. xi. Fitch calls the one approach "rationalist," the other "historically determined."

[24] *What Time Is This Place?*, pp. 38-9.

[25] Joel Garreau, *Edge City* (New York: Doubleday, 1991), p. 58. See Recommended Bibliography.

[26] J. B. Jackson, *The Necessity for Ruins* (Cambridge: MIT Press, 1980), p. 101.

[27] This and the following quote are from Jane Jacobs, *The Death and Life of Great American Cities* (New York: Random House, 1961, 1993), pp. 253-254. See Recommended Bibliography.

1920 - The Salada Building was built in Boston's Back Bay in 1916 for the Salada Tea Company. The Beaux-Arts interior had Chinese decor layered on it, reflective of the wealth brought to Boston by tea brokers like these clerks.

1984 - The Grill 23 and Bar copied the old brokerage chairs, opened up some walls, and took advantage of the high ceilings to raise part of the floor. Columned space invites this kind of space plan variation.

Kentucky, that "started life as a fashionable athletic club, outlived that and became a school, then the stable of a dairy company, then a riding school, then a finishing and dancing school, another athletic club, an artist's studio, a school again, a blacksmith's, a factory, a warehouse, and it is now a flourishing center of the arts." That's twelve incarnations so far, with the building apparently holding its value the whole time. Where does that

ADAPTIVE USE. Robert Campbell wrote about these interior photos: "Recyclings embody a paradox. They work best when the new use doesn't fit the old container too neatly. The slight misfit between old and new—the incongruity of eating your dinner in a brokerage hall—gives such places their special edge and drama.... The best buildings are not those that are cut, like a tailored suit, to fit only one set of functions, but rather those that are strong enough to retain their character as they accommodate different functions over time." (*Cityscapes of Boston*, **pp. 160-161. See Recommended Bibliography.**)

leave design truisms like "Form follows function"? Completely invalidated. The building became more interesting when it left its original function behind. The continuing changes in function turn into a colorful story which becomes valued in its own right. The building succeeds by seeming to fail.

In 1964 an abandoned chocolate factory on the San Francisco waterfront was converted into a shopping complex called Ghirardelli Square. Instantly a major tourist attraction, it became the prototype for adaptive-use commercial projects all over the world. Empty old factories, warehouses, terminals, and enclosed docks suddenly were being bought up by visionary developers. The spaces were big and adaptable, the heavy construction sound, the settings "picturesque," and the city governments under pressure from preservationists to save every big old building, however humble. Urban renewal had found a way to be acceptable—go indoors and upgrade building Services and Space plans instead of replacing whole blocks and buildings. It was so commercially attractive that you could do it with private funds. Adaptive use took off as the mainstream of preservationist activity.

The fact is that obsolete buildings are fun to convert and a delight to use once they're converted. Wouldn't you rather go to school

1979 - The downtown of Akron, Ohio, grew right around the monumental grain silos (1932) of the Quaker Oats Company. People got used to seeing the structure and would have missed it if it was gone.

1990 - The downtown location made the property central enough to support a hotel. That, plus preservationist incentives from the government, turned a bizarre idea into something economically attractive. The hotel is known as the Quaker Hilton.

ADAPTIVE USE opens minds to formerly unthinkable possibilities, such as converting concrete grain silos into a hotel.

1980 - Conversion.

in a former firehouse, have dinner in a converted brick kiln, do your office work in a restored mansion? I don't have to look far for an example: my wife and I live in a 65-foot 1912 tugboat that turned out to be easy to turn into a home. Originality is unavoidable. A building being reconfigured for a foreign new use is filled with novel opportunities and impossible-seeming problems. Both encourage creativity, and you can't brute-force design solutions on an implacably existing building: you have to finesse them. Invention becomes a habit as you proceed.

This is the formal answer to the question, "Why are old buildings more freeing?" They free you by constraining you. Since you don't have to address the appalling vacuum of a blank site, you can put all of your effort and ingenuity into the manageable task of rearranging the relatively small part of the building's mass that people deal with every day—the Services, Space plan, and Stuff. Instead of having to imagine with plans, you can visualize

directly in the existing space. "We'll need another window over there to light this room, which will be much deeper when we take out that wall. And then those stairs will be more central, so we should make them a bit grander." It is much easier to continue than to begin. Less money is needed, as well as less time, and fewer people are involved, so fewer compromises are necessary. And you can do it by stages while using the space. The building already has a story; all you have to do is add the interesting next chapter.

Old buildings are full of details from an earlier time beyond the ken of the current generation of architects and builders. Preservationists insist on keeping early details like that for later appreciation and reinterpretation. As much as possible of the original fabric of the building is to be saved. New work should be potentially reversible. Clem Labine gives an example, "You might, say, add a porch, but you try to do it in a way that doesn't destroy any of the building that's there. You don't knock off anything, and you maybe just bolt it on in a few places, so if somebody else comes along and wants to put it back exactly the way it was, they can."

1912 - MIRENE was built as a gasoline schooner, 36 tons, at North Bend, Oregon, in 1912. Her job was to haul passengers and cargo along the coast and up the rivers of Oregon and not to be daunted by going aground on the bars at the river mouths. (Image reversed for continuity.)

ca. 1945 - Around 1930 the MIRENE was converted to a tugboat with the addition of a new house and a powerful engine. The pilothouse shows the classic flat-fronted "boot-heel" (two-level) shape of Northwest Coast tugs. (Image reversed for continuity.)

ca. 1962 - On the Willamette River near Portland, MIRENE was at her peak as a tug towing log rafts. Her handsome paint job at this point is the one we later imitated.

GOOD OBSOLETE DESIGN FORCES GOOD NEW DESIGN. In converting a 1912 tugboat to a houseboat, my wife and I found that the severe constraints of the tiny, convoluted space freed us. We were pushed toward ideas of an originality and practicality we would never attempt in a conventional house.

1993 - With her stern facing southeast, MIRENE's former engine room invited conversion into a library/living room with French doors to collect the morning sun, openable for summer breezes. The delicate arch of her windows we echoed in the French doors.

1993 - MIRENE's total living space is 450 square feet—bedroom, kitchen, bathroom, and library. The library is 13 feet by 12—156 square feet. The need to conserve space led us to ideas like mounting the bookshelves chest-high above the floor. (The oars and spars go with a rowboat parked outside.)

March 1981. Brand.

1981 - In 1975, in decrepit condition, MIRENE was bought for salvage and sailed to San Francisco Bay. There relieved of her 400-horsepower diesel engine, propeller, and steering wheel, she became a hulk. At this point my wife and I purchased the ruin for $8,000 from a gent named Rupert Pickle. The wood was so rotten, you could grab handfuls of the bulwarks with your bare hands. (Image reversed for continuity.)

23 August 1993. Brand.

1993 - New decks, new bulwarks, an outside ladder to the pilothouse, a sturdy replacement for the plywood box that had replaced the engine room, and a garden of sorts were some of the improvements we made, thanks to shipwright neighbors on the Sausalito waterfront. Wooden boats are so impossible to maintain, they attract artisans with exceptional skill and traditional knowledge.

"Pocket windows"—traditional on working boats in the early 20th century—provide a clever way for the small windows to open by disappearing into the wall below them (right). This inspired us to design a variation where the new windows in the library (left) open by sliding sideways into a space in the wall, thereby not banging in the wind or blocking the narrow deck outside, and easily adjustable.

Our bed in the pilothouse (right) has a view in four directions. The low ceiling let reading lights be attached overhead, and the tongue-and-groove fir walls made it easy to build in a convenience shelf next to the bed. The heater (on its own space-saving shelf) warms the small room quickly and inexpensively.

So-called captain's beds on boats have always featured storage underneath. Our bedroom in the pilothouse is only 11 by 8 1/2 feet and includes a bidet and corner sink. So Elfa shelving went under one end of the bed, while the two-leveled shape of the pilothouse gave room for a hanging closet under the other end.

Originally the pilothouse was two tiny rooms—bridge forward, skipper's quarters aft. Taking the wall out left an awkward step between the two levels, with a head-knockingly low ceiling. The solution was a half-step, learned from New England colonial houses. That left room for what turned out to be the houseboat's most fiendishly clever feature—a little opening that was easily reached from the kitchen below. All the effluvia of married life flow up and down through that opening, plus it serves as intercom and gymnastic cat passage.

The library (left) had voluminous under-deck spaces on the sides, left empty when we got rid of the fuel tanks. They turned out to be the perfect place to hide the hi-fi and CDs and speakers, exercise gear, guest bedding, TV apparatus, library ladder, and the like. Boaty canvas covered them handily. In the bedroom (right) similar under-deck space offered storage for shoes, etc. in restaurant dish-bins (a trick learned from Airstream trailers).

8 October 1993. Brand.

On the same principle, preservationists encourage the leaving of "hidden treasures" for later remodelers to find, since they themselves have so often been thrilled to discover antique kitchen tools stashed in a walled-up original fireplace, or a wallpaperer's proud signature and date on the plaster under the earliest layer of wallpaper, or old bottles in the floors and newspapers in the walls, or a child's secret hiding place for toys. Remodeling inevitably reveals information from the past and leaves information for the future. Workers sometimes leave mementos and messages for each other that the owners know nothing about.[28]

What about previous improvements and additions to an old building that might be seen as violating its original intention and integrity? Current preservationist doctrine says to respect them; they are a valid part of the building's history. Clem Labine takes a pragmatic approach: "If it works, and it's built well, and it's a nice example of whatever it was intended to be, and it's not getting in the way, then I say leave it." When a surface is richly layered, leave the existing layers and add another. An historic hotel in Hollywood was remodeling the ballroom and found behind the ornate ceiling six previous ceilings, each in a different outmoded style. It was a high enough room; they added the new ceiling without disturbing the others. The work was cheaper that way, too.

Some kinds of buildings are impossible even for ardent preservationists to find new uses for, usually because they are too specialized or too large. A big institutional prison is good for only its original purpose, and it was constructed to discourage change by the occupants. Once it's obsolete, all you can do is leave it as an historic building, like Alcatraz in California, or tear it down.

---

[28]　Tracy Kidder tells of a message found in a can above a porch in an old house in Delaware: "William W. Rose in closed this on November 25th, 1850. On the night of the 23rd of the same month Josiah Ridegeways dwelling house and storehouse and wheelwright shop was burnt down on the corner And this frame suffered from the same fier. With difficulty it was savd. I in close this that when this house is wore out the repairer will know how long it was bilt. James Stevens had this house bilt. He was a Blacksmith by trad the best ax man of that time. William W. Rose was the bilder." Tracy Kidder, *House* (Boston: Houghton Mifflin, 1985), p. 140.

ca. 1985 and ca. 1980 - St. Michael's Church (1858) in Derby was declared redundant in 1977—there were seven other churches within walking distance.

CHURCHES ARE DIFFICULT to adapt, as England knows only too well, having 12,000 empty medieval parish churches plus many thousands of more recent construction, most of them "redundant" because only 14 percent of Britons go to church on Sundays any more. The conversion problem is similar to that of theaters, except that the flat floors and side aisles offer some purchase for subdividing the space into offices, meeting rooms, dance studios, restaurants.

Theaters have a huge, oddly segmented space with few horizontal surfaces, difficult to subdivide except into movie multiplexes. Many stand empty; a few become supermarkets because of convenient location.

Other kinds of buildings seem to be infinitely convertible. Houses go on forever if the property value lets them, and not just as homes. Old townhouses with good addresses become professional offices. The ones with modest-to-poor addresses are subdivided into rental flats, with a shop on the ground floor. Country mansions become conference retreats and schools. Houses are the one species of building most thoroughly co-evolved with human use, and that congeniality carries over, whatever they are used for later.

Warehouses and factories that were built between 1860 and 1930

1982 and ca. 1985 - Architect Derek Latham was persuaded to convert the building to three floors of design offices, with his own office occupying the middle floor. The conversion cost £120,000, whereas a new building of comparable space would have cost £200,000.

are endlessly adaptable. They are broad, raw space—clear-spanned or widely columned, with good natural illumination and ventilation and high ceilings of 12 to 18 feet. The floors, built strong enough for storage or to hold heavy machinery, can handle any new use. Their heavy timbers and exposed brick appeal to the modern eye. Architectural ornament, if any, is likely to be modest and therefore appreciated. The buildings are honest, generic, sound, and common. They welcome any use from corporate headquarters to live/work studios. The modern equivalents of these buildings—the windowless tilt-up concrete structures out on the edge of town—have nothing like the adaptivity of the old brick warehouses.

Another trait that invites longevity is strangeness. Almost any sufficiently odd building that has a modicum of functionality (hollow elephants seldom make it) will attract supportive community bemusement and a sequence of creative occupants. Fantasy draws people to places like the Bridge House in Ambleside, Cumbria, England. It is an adorably tiny stone hut built spanning a small brook as a sort of folly by the owners of the local manor in the 16th century. With just two rooms, one for each story, it looks like a little tower on its stone arch over the water. Its uses over the centuries have included counting house, tea-room, weaving shop, home (for a family of eight), cobbler's shop, gift shop, and most recently, information center for The National Trust. The building seems assured of at least another 400 years. Who could tear down such a durable delight?

Adaptive use is the destiny of most buildings, but the subject is not taught in architectural schools. Any kind of remodeling skills are avoided in the schools because they seem so unheroic, and the prospect of remodeling or rehabilitation happening later to one's new building is even more taboo. Predictably paralyzed buildings are the result. But suppose preservationists taught some of the design courses for architects, developers, and planners. The subject would not be how to make new buildings look like old ones. It would be: how to design new buildings that will endear themselves to preservationists sixty years from now. Take all that a century of sophisticated building preservation has learned about materials, space-planning, scale, mutability, adaptivity, functional tradition, functional originality, and sheer flash, and apply it to new construction.

No preservationist I talked to was interested in this idea, but not until it is occurring routinely will the preservationist program be complete. The wisdom acquired looking backward must be translated into wisdom looking forward. While working toward that day, preservationists can congratulate themselves for the sweeping revolution they set in motion with one simple change to the economics of real estate. It used to be that old buildings were universally understood to be less valuable than new. Now it is almost universally understood that old buildings are more valuable than new.

Under pressure from preservationists and environmentalists, economists have come up with a new term to reflect the newly understood value. Old buildings, like old forests, are appraised as *intergenerational equity.*

1956

1959

CHAPTER 8

The Romance of Maintenance

A FORTUNE in comic books and baseball cards—over a million dollars' worth of collectible treasure—was stored in an Emeryville, California, warehouse. Collector/entrepreneur Robert Beerbohm had thirty employees dealing the collection to a hundred companies worldwide. One night in 1986 it rained harder than usual. The flat warehouse roof ponded up and leaked torrentially—it had a known drainage problem, but maintenance had been deferred by the building owners. Water two feet deep turned the comics and cards to a smelly pulp that had to be trucked off as landfill. The insurance company cited a technicality and refused to cover the loss. His business gone, Beerbohm had a heart attack and a nervous breakdown which left him physically numb for two years.

No wonder people get in a permanent state of denial about the need for building maintenance. It is all about negatives, never about rewards. Doing it is a pain. Not doing it can be catastrophic. A constant draining expense, it never makes money. You could say it does save money in the long run, but even that is a negative because you never see the saving in any accountable way. (If Beerbohm's warehouse roof had been fixed, he would not have noticed.) When, after months or years of nagging, you finally do the work—refinish the floor, hire the roofers, replace the damned furnace—you have nothing new and positive, just a negated negative. The problem that needed fixing turned into an even worse problem during the chaos of repair, and then it went away. Even the Bible is on your case for waiting so long: "By much slouthfulness the building decayeth." (Ecclesiastes, X, 18.)

Yet the issue is core and absolute: no maintenance, no building.

1961 - Originally built around 1790, Gibson Wright's Mill on the upper Eastern Shore of Maryland was timber frame on a brick foundation. In the early 19th century, some of its beaded clapboards were replaced by beveled clapboards, and the windows were widened.

1965 - Building historian H. Chandlee Foreman recorded the demise of Wright's Mill and published the pictures in his *Old Buildings, Gardens, and Furniture in Tidewater Maryland* (Cambridge MD: Tidewater, 1967).

WHO BUILDS IN WOOD builds a shack—adaptable now, gone soon. No other material is so easy to work, and none is so vulnerable to neglect, except maybe adobe. Once the roof and windows of Wright's Mill were open, even its sturdy timber frame construction could not save it from the effects of constant moisture. To insects and fungus, wet wood is food.

And that's what usually happens. Every building is potentially immortal, but very few last half the life of a human. Preservationists are so adamant on the subject that the motto of their department at the US National Park Service declares: "Preservation IS maintenance." John Ruskin himself, the founder of anti-scrape preservation, intoned, "Take proper care of your monuments, and you will not need to restore them. A few sheets of lead put in time upon the roof, a few dead leaves and sticks swept in time out of a water-course, will save both roof and walls from ruin. Watch an old building with an anxious care; guard it as

best you may, and at *any* cost, from every influence of dilapidation."[1]

According to realtor hearsay, only a third of American houses are well-maintained, and that fraction has been getting smaller as people's work hours increase. The apparent shortening of American attention span is no help either, since maintenance problems come on with such a slow relentlessness, they conveniently elude perception. "Is the porch starting to sag?" "Didn't it always sag?"

---

1 John Ruskin, *The Seven Lamps of Architecture* (New York: Dover, 1849, 1880, 1989), p. 196. Ruskin knew about roofs. He personally crawled over countless buildings in Venice and elsewhere.

The sequence of effects of deterioration on ordinary buildings has never been formally studied—a curious lapse, considering the massive capital loss involved—but some rules of thumb have emerged. Due to deterioration and obsolescence, a building's capital value (and the rent it can charge) about halves by twenty years after construction. Most buildings you can expect to require complete refurbishing from eleven to twenty-five years after construction.[2] The rule of thumb about abandonment is simple: if repairs will cost half of the value of the building, don't bother. This is the point at which owners either demolish the place (often leaving a vacant lot, which is considered to be more salable than a shabby building, to the despair of city planners and preservationists), or they burn it for insurance, or they let it stand empty.

An empty building rots fast and attracts trouble. Once it is left unheated and unventilated, any moisture that gets in immediately begins causing serious damage, with no one around to notice or worry. Vandals smash the windows, letting in more rain, and they trash the interiors. Now considered an eyesore and a hazard by the community, the building won't be allowed to continue much longer.

Since the downward spiral of dilapidation can accelerate so quickly, the trick is to keep a building from entering the spiral at all. Two methods are supposedly standard, but both are in practice somewhat rare. One is "preventive maintenance"—routinely servicing materials and systems in the building *before* they fail, thereby saving considerable expense and greatly extending the life of the building. The other is designing and constructing the building in such a way that it doesn't need a lot of maintenance. Both are unpopular. Solid construction? Expensive! Preventive maintenance? BORing.

Building maintenance has little status with architects. They see the people who do the maintaining as blue-collar illiterates and the process of upkeep as trivial, not a part of design concerns. So they will use gaudy stainless steel cladding unaware that the stuff corrodes and falls off. The prize-winning Queens Building at De

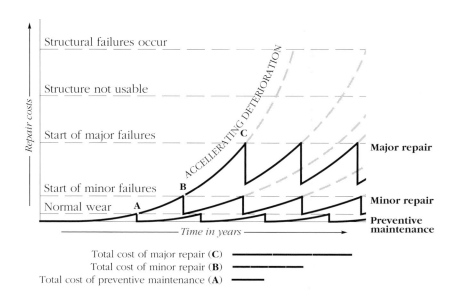

PREVENTIVE MAINTENANCE (bottom line) not only costs markedly less in aggregate than repairing buildings failures, it reduces human wear and tear. A building whose systems are always breaking or threatening to break is depressing to the occupants, and that brings on another dimension of expense. This diagram is adapted from *Preventive Maintenance of Buildings* (New York: Van Nostrand Reinhold, 1991), p. 3.

Montford University in Leicester, England, requires a 60-foot scaffold just to change light bulbs, and its main power room is so undersized, moving in any equipment means turning off the electricity for the building. But even ordinary office buildings suffer from the lapse. According to *The Occupier's View*, the survey of fifty-eight new business buildings near London, "a staggering one-fifth of the sample said that the need to clean their windows had not even been considered during the design and construction of the building."[3] Also lighting fixtures in the grand lobbies were unreachable for lamp replacement, and internal drains from the flat roofs had no access hatches for inspection and cleaning.

Incompetent design often is matched by hurried, shoddy construction, which can be concealed or can get by on being just good enough, just long enough. A building scientist I talked to,

MAINTENANCE NIGHTMARE. Pompidou Centre (1979) is considered a landmark in architecural history, and it is a major tourist attraction, competing with the Eiffel Tower. Where the Eiffel Tower (1889) exposed its structure in an elegant and monumental way, Pompidou Centre tries to do the same with its services. But iron structure can withstand the elements much better than brightly painted ducts and pipes. The Eiffel Tower's lasting message to architecture is: exposed structure can be gorgeous. Pompidou's lasting message: never expose services.

December 1989. Brand.

1989 - The services facade of Pompidou Centre in Paris.

Terry Brennan, observed that builders now don't take the time to test and tune a new building. There's no commissioning period such as ships get. Frequently, things like sensors and ventilation systems are installed but never even plugged in. Money and time is invariably tight at the end of the project, so "details" get skimped or overlooked. Brennan concluded, "The older a building, the more likely it is to be right. Since the 1970s, buildings don't work."

The trend in construction during this century has been toward ever-lighter framing, at least in America, with the result that buildings look and feel increasingly like movie sets: impressive to the eye, flimsy to the touch, and incapable of aging well. Only the buildings that are aggressive about energy conservation have countered the trend, with their thicker, well-insulated walls and

carefully airtight construction. European houses, one hears, are still being built robustly, often of masonry. They cost 50 percent more than American homes to build, but they make that up in saved maintenance expenses over the first fifteen years, after which they are ever more economical than their American equivalents. European families think in generations while Americans are still trying to master decades.

Our nation of immigrants may never get around to thinking in

---

[2] Roger Flanagan, et al., *Life Cycle Costing* (Oxford: BSP, 1989), pp. 44-45. Shopping centers cycle in eleven years, warehouses in twenty-five years, with other commercial uses in between.

[3] Vail-Williams, *The Occupier's View* (Fareham: Vail-Williams, 1990), p. 185. See Recommended Bibliography.

generations, but taking the long-term view is on the increase. Preservationists, and perhaps the maturing forces of history, are Europeanizing America. City identities are solidifying around the existing inventory of buildings. Colby Everdell, an architect with the multinational construction firm Bechtel, figures that his company needs to think now about constructing hundred-year buildings in cities instead of the usual thirty-year ones, "because they don't let you tear down buildings any more."

The longer that buildings are expected to last, the more you can expect maintenance and other running costs to overwhelm the initial capital costs of construction, and the more inclined owners will be to invest in better construction so they can spend less on maintenance. If you can't count on discarding a building, and you can save $300,000 over a few years by spending $50,000 now, the investment is worth it. In a building-conservative economy, low-maintenance or well-maintained buildings are bound to hold their rental and sale value better than the usual frapped-out throw-away structures.

So then, what makes buildings last well? And what does them in?

The root of all evil is water. It dissolves buildings. Water is elixir to unwelcome life such as rot and insects. Water, the universal solvent, makes chemical reactions happen every place you don't want them. It consumes wood, erodes masonry, corrodes metals, peels paint, expands destructively when it freezes, and permeates everywhere when it evaporates. It warps, swells, discolors, rusts, loosens, mildews, and stinks. Leon Battista Alberti, inspirer of architects, lamented in the 15th century, "Rain is always prepared to wreak mischief, and never fails to exploit even the least opening to do some harm: by its subtlety it infiltrates, by softening it corrupts, and by its persistence it undermines the whole strength of the building, until it eventually brings ruin and destruction on the entire work."[4]

Rain is only the most obvious source of the problem. More pernicious now in most buildings is internally generated water vapor. Every year since the energy crisis of 1973, buildings have been made more airtight and better-insulated to save on energy costs. But keeping in the nice warm air (or the nice cool air in hot seasons) meant also keeping in all the moisture that humans, kitchens, and bathrooms constantly exhale.[5] Being a gas, water vapor can penetrate everywhere in search of a cool surface to condense on. When it gets into a wall and condenses there, it soaks insulation, corrodes anything metal (nails, screws, bolts, wire lath, masonry ties), and flows down to the floor plate to rot the building's basic structure—all hidden from view. The owner doesn't learn of the problem until a hoped-for buyer's building inspector reports, "Oh by the way, your walls are cheese."[6]

Bathrooms these days are ventilated with electric fans not to remove noxious odors but to evacuate the even more noxious water vapor. "Houses seem to deteriorate from the bathrooms out," observes home repair specialist David Owen. "A neglected bathroom can metastasize far into a house, carrying destruction with it."[7] Because of water, houses deteriorate most from the bottom up and the top down. Damage comes from below thanks to what the British call, knowledgeably, "damp." Moisture from

---

[4] Leon Battista Alberti; Joseph Rykwert, et al., translators, *On the Art of Building in Ten Books* (Cambridge, MIT Press, 1988), p. 27.

[5] Human lungs respire eight to twelve pounds of water every twenty-four hours. Cooking on a gas stove for a family of four puts out 5 pounds a day. Every shower generates half a pound. Sedway Cooke Association, *Retrofit Right* (City of Oakland, 1983), p. 58.

[6] The wall-condensation problem led to a great many healthy walls suddenly going bad in the 1970s, when buildings were first being retrofitted to save energy. The cause, which took a few years to figure out, was that the vapor barrier was being installed in the wrong place. If you put the vapor barrier against the outside wall in a cold climate, it will condense moisture destructively. But in a hot humid climate, if you put the vapor barrier against the air-conditioning-cooled *inside* wall, then moisture from outside collects there. Local adaptations were gradually figured out, including having negative air pressure (such as with open fireplace chimneys) in chilly climes and positive air pressure (such as with air conditioning) in hot moist regions. However much technology we may add, buildings are still utterly creatures of their climate.

[7] David Owen, *The Walls Around Us* (New York: Random, 1991), p. 235. See Recommended Bibliography.

[8] *The Occupier's View*, pp. 105-106 and 181.

[9] Built-up roofing was increasingly replaced in the 1980s and after by "single-ply" roofing made of one layer of plastic, relying even more on the perfect membrane theory. Their durability and flexibility is still in question. I wouldn't trust them.

the ground rises by capillary action into foundations and seeps into cellars and treats a building like a very large lump of sugar.

Still, top-down is the main event. Water can rise only a few feet, but it can descend and branch out indefinitely. A building's most important organ of health is its roof. Roof effectiveness is determined most by its pitch and shape, next by its detailing, next by its materials, and last by its look, which is irrelevant. Architects who have indulged Post-Modernism's penchant for pitched roofs—all those triangular gables, even on top of skyscrapers—have been startled to discover that these roofs work better than what they're used to. That's because what three generations of architects are used to is flat roofs. Flat roofs are convenient and safe to work on, and they disappear behind parapets so that all the junky-looking vents and coolers and whatnot of modern buildings are out of sight. Modernist architects bought into flat roofs as a standard idiom because they were new and radical (back then). They were consistent with the idea of building-as-machine-for-living, and they were seen as ultimately flexible—you could put a stairwell and skylight any place you wanted. There was just one problem, which was overlooked for most of the 20th century.

"Flat roofs always leak," said an estates manager interviewed by the researchers for *The Occupier's View.* The researchers observed that even those few occupiers (22 percent of the sample) "who had expressed a preference for modern glass boxes seemed to be resigned to the fact that part of the price of a modern building with a flat roof was the likelihood (or virtual certainty) of roof leaks." The overwhelming majority of the occupants of the fifty-eight modern business buildings—78 percent of them—said that after their experience they would prefer to be in a building with "traditional pitched tile roof and brick construction."[8]

In theory, flat roofs can work if the covering surface is a perfect membrane. In practice they fail because the membrane isn't perfect. The technology of the standard "built-up roof" (BUR) has been around for a hundred years, utilizing four or more alternating layers of roofing felt and asphalt, with a mineral layer such as gravel on top to absorb the sun's destructive rays. Problems arise during construction when the weather is cold or wet and from poorly installed flashing. Further problems come later when people walk on the roof and when they add new service equipment penetrations which violate the membrane. Routine failures include erosion (gravel scoured away by the wind), alligatoring (sun oxidation), blisters (from expanding moisture in the roof layers), splits, ridges, punctures, and fish mouths (where seams between the felts open and pooch up).[9]

The worst of it is, when water comes through a flat roof, *you can't tell where the leak is* because the water travels great distances hidden in the roof, ceilings, and walls before visible damage shows up. (In a pitched roof, damage is limited to the area, usually visible or traceable, just below the leak.) The only solution is best-guess repairs, which generally fail, followed by a whole new layer on the roof. But that can become part of the problem as successive heavy layers weigh down the roof structure and make some areas concave. Once water can pond, leaks are inevitable and untreatable. You have to rebuild the roof. The easiest solution, and a common one, is just put in a pitched roof over the old flat one.

It seems obvious, but people keep having to rediscover it: pitched roofs shed water well. Also, they provide a volume of air under the roof that absorbs radiant heat and water vapor and vents them to the outside. That space can be used for storage and for service equipment that otherwise would be outside in the weather or down in the building bothering people. And pitched roofs can have eaves that overhang enough to protect the walls from rain and sun and even keep water from collecting in the ground around the foundation.

The simpler the roof, the better. Chimneys, dormers, valleys, skylights, widow's walks, and other complications all invite problems with flashing, and they make reroofing difficult. Built-in gutters look nice, but they often wind up leaking into the walls.

Much better are the standard clunky hanging metal gutters. They need to be cleaned frequently (especially in the fall and winter) and downspouted to discharge rain well away from the foundation. Hand-crafted skylights always leak; factory-built almost never do. A seldom-utilized but highly important roof element is color, the lighter the better. With a white or silvered roof reflecting away the sun's heat and destructive ultraviolet rays, the building will be far more comfortable and energy-efficient, and the life of the roof material will be doubled.

Representing some 70 percent of a building's exposure, the roof has to take extreme punishment—from rain, snow, and ice, from freezing and frying (and the contracting and expanding that go with them), from wind, from chemicals in the air, and from constant molecular breakdown by the sun's ultraviolet rays. The side of a pitched roof facing the weather (sun or wind, whichever's worse locally) will need new roofing in half the time of the protected side. If ever you want to do a building a favor, buy it a new hat.

A hat made of what? The choices are: wood shingle, asphalt composition shingle, built-up roof, single-ply membrane, lead, tile, slate, and metal. Wood shingle looks nicely weathered in just a year, but it only lasts about fifteen years, if fire doesn't get it first.[10] George Washington had to replace Mount Vernon's shingles six times. Composition shingle is cheap and comes in colors, but it lasts only fifteen to twenty years. The single-ply membranes are still too new to gauge their longevity, and built-up

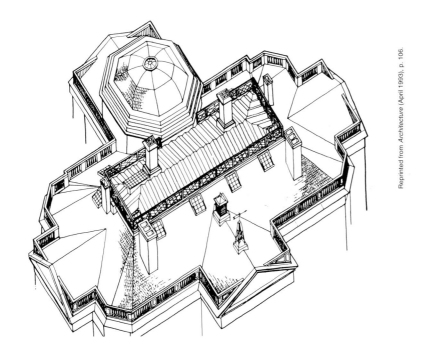

Reprinted from *Architecture* (April 1993), p. 106.

The 1992 restorers of Monticello's roof—Mesick Cohen Waite—made this axonometric view of what they had to deal with. The restorers had good records to work from, thanks to Jefferson's voluminous correspondance with his builders while he was away for years serving the new American government.

JEFFERSON'S COMPLEX ROOF may have been what drove him to all his experiments with roofing materials. "With thirteen skylights, six chimneys, an elaborate balustrade, a 400-foot entablature, and a dome—the first ever to crown an American house—Monticello is considered to represent the most complex roofscape of any house in early 19th-century America." (Marc S. Harriman, "Jeffersonian Invention," *Architecture*, April 1993, p. 105.) Restoring the roof was a $1 million job. The restorers used tin-plated stainless steel rather than Jefferson's tin-plated wrought iron (imported from Wales), and the shingles were lapped by four inches this time, instead of Jefferson's two inches, which had apparently let in wind-driven rain.

roofs are known for short lifespans of ten to twenty years.

The 100-year-plus materials are lead, tile, slate, and metal. When lead finally corrodes after a century or so, it needs to be completely replaced. Tile and slate are heavy, expensive, and sometimes breakable, but they are fireproof and beautiful, and they will last the life of most buildings (often much longer, since they can be recycled). New concrete tiles are not as attractive as traditional clay tiles—a 12,000-year-old technology—but they cost less. Slates are soulful. However, they don't hold up in sunny climates quite as well as tile (ultraviolet rots them), they need steeper pitches to reduce moisture damage, and they require stainless steel or copper nails if you want the fasteners to last as long as the slates.

November 1991. Brand.

WASHINGTON'S WOOD SHINGLES for Mount Vernon took such punishment from sun and weather that they were turned over and reused.

The museum at Mount Vernon has interesting material on the construction history of the house, including this early shingle. Like Jefferson, Washington left abundant (if often angry) correspondence about what he wanted for the building because he expanded the house while battling the British.

Metal roofs have become tremendously popular since architects began getting sued for leaks. The best of all is standing-seam terne-coated stainless steel or copper. It is light, nonflammable, moderately priced, good looking, nearly maintenance-free, and waterproof (it also sheds snow, branches, and prowlers). In *The Low-Maintenance House*, Gene Logsdon reports, "Every roofer I ask says that metal roofs today are the best buy for the money of *any* kind of roofing."[11] Len Lewandowski concludes in *Preventive Maintenance of Buildings*, "The standing seam roof offers the lightest weight, lowest maintenance, and most cost-effective roofing solution available today."[12] Survivors of hurricanes in the American southeast say that metal roofs should be fastened with screws rather than nails—stainless steel of course.

After the roof, the most vulnerable part of a building's exterior is the windows. Like people, buildings would have far fewer upkeep problems if they had no orifices. Water dampens and sun toasts the horizontal surfaces, and decay blooms in the cavities and crevices. Condensation collects on the inside of the glass. The moving parts undergo wear and tear. And even more than the rest of the building's skin, windows obsolesce quickly from fashion swerve and technology advance. Most won't last twenty years.

When it comes to walls, one of the great cautionary tales of maintenance is the siding question. Exasperated home owners are offered a shortcut: just put up aluminum or vinyl siding and quit worrying forever about peeling paint and decayed wood and

[10] Following the 1991 fire in Oakland, California, that destroyed thousands of homes, comparative studies were made of roof flammability. Buildings with wood-shingled roofs were found to be 50 percent more likely to catch fire than any other kind. They are now totally banned in Oakland. David Moffat, "Planning Fire-Safe Design," *Architecture* (Sept. 1992), p. 112.

[11] Gene Logsdon, *The Low-Maintenance House* (Emmaus, PA: Rodale, 1987), p. 68. See Recommended Bibliography.

[12] Len Lewandowski, "Exposed Metal Roof Systems," in Raymond Matulionis and Joan Freitag, eds., *Preventive Maintenance of Buildings* (New York: Van Nostrand Reinhold, 1991), p. 122. Lewandowski mentions that by the late 1980s "57 percent of all new, low-rise, nonresidential buildings" were getting metal roofs, and many older buildings were being retrofitted with metal. The minimum roof slope for metal is 3 inches per foot.

all the rest of it. Aluminum eventually dents, and its paint can scratch off, but vinyl (available since 1963) has neither disadvantage. At the cost of three paint jobs, put it up and your problems go away, right? Wrong. Where your problems go is out of sight. Vinyl siding is a vapor barrier chilled by outside cold. Any moisture behind it, whether from leaks or condensed house humidity, is trapped to do its damage invisibly for years. The damage can be structural.

The question is this: do you want a material that looks bad before it acts bad, like shingles or clapboard, or one that acts bad long before it looks bad, like vinyl siding? A whole philosophy of maintenance falls one way or the other with the answer. What you want in materials is a quality of forgivingness. Shingles and clapboard expand and contract comfortably with temperature extremes, they let water vapor through, they show you when they're getting worn, and they're easy to replace piecemeal. The same is true of British tile-hanging on exterior walls, a weatherproofing practice worth importing to the US. Remodelers love shingled and tile-hung walls because changes are so easy to make and then hide.

The attraction of traditional materials such as shingles and clapboard is more than just aesthetic. Their whole use cycle is a highly evolved system of trade skills, reliable supply sources and routes, generations-deep familiarity, and even a market for reuse of durable materials such as slates, tiles, bricks, and timbers. The problems of traditional materials are thoroughly understood, and the solutions are equally well known. Maintenance is no mystery. In some cases maintenance can be a matter of steady improvement, as with the now-unfashionable use of whitewash on masonry or stucco walls. In the days when medieval castle walls were routinely brightened inside with fresh whitewash, it was said that the whitewash "fed the stonework." It did, and it does so to this day on the dazzling rubble-and-stucco buildings of the Greek islands which are required by local law to get a new coat of whitewash annually. The lime or chalk in the whitewash fills hairline cracks before they expand and helps keep water out.

January 1969 - Victorian detailing adorned this 1879 three-decker at 48 Berkshire Street in Cambridgeport, Massachusetts.

Robert Bell Rettig. Cambridge Historical Commission. Neg. no. 193/16A.

December 1969 - After aluminum siding went on, the Cambridge Historical Commission documented (and presumably deplored) the difference.

E. Jacoby. Cambridge Historical Commission. Neg. no. 270/32A.

ALUMINUM SIDING hides sins and gains only the illusion of lower maintenance. Clem Labine, founder of the *Old House Journal*, condemns it because it is irreversible: "It's the difference between taking a Queen Anne house and painting it purple versus putting aluminum siding on it. Painting it purple is a totally reversible act. If you put aluminum siding on the same house, the installer may have to take an ax and chop off ornament and things that don't fit tidily under the siding. If someone later goes to all the trouble to rip off the siding, they don't have the original house there. It's been considerably mauled."

Traditional materials are considered aesthetic partly because of their rich texture but even more because they are thought to age attractively. They have a time dimension. Artisans even learn how to ripen them artificially, as with the trick of using cow urine on shiny copper sheeting to make it match the older part of a copper roof. Recent materials such as aluminum or exposed concrete are said to age ugly, but that may be a function of their youth. Just as any building over a hundred years old is declared to be beautiful, any venerable material is given the same courtesy.

New materials are unproven, by definition. Like most experiments, they tend to fail. If the experiment is the whole exterior of a highly visible building, they fail big. "Curtain walling systems seem to be alarmingly prone to leaks," reported the surveyors for *The Occupier's View*.[13] (The interviewees added that floor-to-ceiling glass made people uncomfortable, and the offices looked messy from outside. The windows seemed to be advertising the firms' ankles and wastebaskets.) "The reliance on an exposed single barrier of sealant to prevent water penetration

---

[13] *The Occupier's View*, p. 90.

into a wall is the single most common source of current building facade problems," summarizes *Architecture* magazine, adding that by 1980 some 33 percent of lawsuits against architects were for facade failures.

"Reliance on a single barrier" is the key defect of new materials. That is what made geodesic domes so leaky. Intelligently designed exterior walls employ what is poetically called "rainscreen" design, which assumes that water will occasionally get through the exterior layer, but it is intercepted and quickly returned to the outside. The multiple layers of shingles, clapboards, and cavity-wall masonry all work that way. Redundancy of function is always more reliable than attempts at perfection, which time treats cruelly. Consider Australia's pride:

> The Sydney Opera House, one of the most memorable buildings built during this century, was finished in 1973 at a cost of $120 million, with a cost overrun of roughly 1,700 percent. The magnificent roof shells, and the building is all roof shells, were designed to last 300 years and may last even longer. Yet, the waterproof joints between them were sealed with mastics that had a projected life of 12 years with no sensible provision made for inspection, maintenance, or repair. In 1989 it was estimated the Opera House would cost $500 million to replace.[14]

Preservationists despair at the prospect of trying to preserve modern buildings. The materials are often one-offs that would take a whole industry to reproduce, and the failures can be massive. The 80-story Amoco building (1974) in Chicago was originally faced with 1-1/4-inch-thick panels of prime Carrara marble, which soon dished and distorted because it was cut too thin. Replacing the 43,000 panels with 2-inch thick granite is taking three years and $80 million.[15]

As for wood, redolent with tradition, it is the best of materials from the standpoint of adaptability and one of the worst in terms of maintenance. It is fairly cheap, made of a renewable resource (theoretically), easy to work, and it can be extraordinarily beautiful. But it always wants to absorb moisture, and wherever the water content gets over 21 percent, the wood turns into habitat and food for fungus, termites, ants, beetles, bees, borers, and other wildlife. "What holds up that house?" one cynical carpenter asked me rhetorically, gesturing at a nearby standard American stick-built home. "Faith, habit, and the dead bodies of termites, same as all the houses around here." Who builds in wood builds a shack, adaptable now, gone soon.

The exception is timber-framed buildings, because the wood structure is protected from the weather, it is massive, and it is exposed. Air and eyeballs can get at it to keep it dry and inspected. "According to government statistics," reports Gene Logsdon, "the average life of a conventionally built stud house is about 75 years. The life of a timber frame is at least 300 years, and some over 1,000 years old survive."[16] Even if the building has to be demolished for economic reasons, the timbers have salvage value in the flourishing old-timbers market, because they can be easily disassembled and reassembled.

Bricks and stone are so much the opposite of wood that the conversion of northern European house construction to masonry in the 16th and 17th centuries is remembered as an historic event—"the Great Rebuilding" in England, "the Victory of Stone Over Wood" in France.[17] Brick is a superlative building material, the product of 8,000 years of experience in firing clay into modular units that can be mortared together and stacked by hand into unreinforced structures as high as sixteen stories.[18] It is

---

[14] Forrest Wilson in *Blueprints* (Winter 1991), the newsletter of the National Building Museum in Washington, DC.

[15] *Architecture* (Feb. 1991), p. 79.

[16] Gene Logsdon, *The Low-Maintenance House* (Emmaus, PA: Rodale, 1987), p. 7.

[17] J. B. Jackson, *Discovering the Vernacular Landscape* (New Haven: Yale, 1984), p. 95.

18 The Monadnock Building, one of the world's first skyscrapers with its sixteen floors, was built of brick in Chicago in 1892. It still stands, a monument to the "Chicago School" highrise revolution.

August 1888. Charles B. Webster. Society for the Preservation of New England Antiquities. Neg. no. 15797-B.

June 1936. Thomas T. Waterman. Library of Congress. HABS Neg. no. 305675 MASS 11-DED 1-7.

1888 - In 1636 Jonathan Fayerbanke built a vernacular English house in the new Massachusetts colony for his wife and six children. A huge central chimney (6 by 10 feet) made of 40,000 English bricks brought over as ship's ballast served three fireplaces and helped anchor the structure against the centuries. Its steep roof was originally roofed with thatch, later with slate (1725), later with wood shingles. In 1648, when the eldest son, John, started a family, Jonathan had part of a gambrel-roofed barn (left) dragged over and attached to the original house, which was extended six feet at the same time. That must have worked well, because in 1654 the rest of the barn was attached at the other end (right), perhaps for hired help. (Building historians doubt this account from family tradition, saying that gambrel roofs didn't show up until the late 18th century, and these were probably built in place.) A further extension was a lean-to shed (middle foreground) added in 1668. After suffering from snow load, it had its roof steepened a few years later—a form which became generalized as the standard New England saltbox. The house had grown from four rooms to fourteen, plus attic and cellars.

1936 - As generations of Fairbankses poured through the house, the only exterior addition was a connected outhouse (1720). Inside, plaster came to some of the walls in 1740, wallpaper in 1830. The cavernous original fireplaces were partially filled in to be more efficient in 1780, and a beehive oven was added. Closets, after they were no longer taxed as rooms, were added in 1870. Three maiden Fairbanks daughters occupied the house quietly through the turbulence of the 19th century and were satisfied to keep the place just as it was.

"A PIECE OF THE TRUE CROSS" is how building historians refer to the Fairbanks House in Dedham, Massachusetts. Built in 1636, it is the oldest wood frame house in America. To get that kind of longevity in a wood structure took a combination of solid timber-frame construction, an unusually massive central chimney of brick, centuries of diligent care by the Fairbanks family, and luck. Another preserving factor might have been the way its parts were kept separate as it grew—the three major living areas had independent entrances and no connections upstairs or in the cellars. As many as three children in each generation could lead separate adult lives in the building, which helped ensure continuity of owner- ship and family pride through eight generations.

2 May 1991. Brand.

1991 - In 1903 the last Fairbanks moved out, and the place was incorporated as a house museum by the family. When I visited, Lloyd Fairbanks here gave the tour. He is the great great great great great great grandson of Jonathan, still tending to the timber-frame house that served eight generations of Fairbankses—eighty-four people, thirteen families. Full of original furniture and antiques, it is one of the realest house museums in America.

8 August 1991. Brand.

"A NO-MAINTENANCE BUILDING" is how the owner describes this tile-roofed brick house in Oxfordshire, near Faringdon. That is, it was after he had coated the two most weatherly walls with clear silicon seal and had one whole course of bricks near the ground damp-proofed by silicon injection. The wood window frames now need a coat of paint every three years, and that's about it for exterior work.

Built cheap but solid around 1900, the building was designed to house two farm laborer families. When Philip Duff bought the building in 1985, one side had been vacant and open to the weather for three years, the other side for eight years. The interior was a mess, but the building was sound. Masonry buildings can endure neglect that would destroy other structures.

Seamlessly joining and expanding the original duplex was made relatively easy by the bricks. The left hand porch was removed and filled in, while the right hand one was expanded into a wind-protected mud room. On the left end, Duff lengthened the whole building six feet with brick matched to the original. (To continue the original English bond—alternating stretcher and header courses—he decided against putting in a cavity wall.) Roof tiles recycled from other local buildings of the period made a perfect match with what was there before. The duplex made it tolerable to occupy the building during the five years of its gradual rehabilitation. Duff could live in each side while the other was being worked on, before finally chiseling through the eighteen-inch party wall.

completely fireproof. London had no more Great Fires after rebuilding in brick in the 1670s, and commercial building owners in the 1990s still favor brick because they more than make up for its extra expense with saved insurance costs. As of 1989, brick was the most popular of all exterior cladding materials in the US for nonresidential construction—31 percent of the market.[19] Maintenance is minimal. All brick walls need is to be repointed (the outer 3/4 inch of the mortar joints replaced) every sixty to one hundred years.

"Bricks are heavenly," says contractor Matisse Enzer, "because they require relatively little technology to create, build with, and *modify*. Bricks allow a wonderful variety of patterns and degrees of softness-hardness, permanence-temporariness. Most of all, they are *intuitively obvious*." Bricks, more than any other material, look like they were made to fit the human hand. With dimensions of 8 inches by 4 inches by 2-2/3 inches, one brick long equals two bricks wide or three bricks high (including mortar joints), so a wide variety of bond or decorative patterns is possible. And they are manufactured in a boggling variety of colors and textures. At the Building Centre on Store Street in London is a "brick library" displaying thousands of commercially available brick samples— Sandcreased, Coarse Sandfaced, Combed, Rock Faced Calcium Silicate, etc. Architects love brick for its rich palette of color, texture, and shape. Brick easily handles otherwise difficult shapes such as curved walls, and masons are glad to show off their craft with window arches, string courses, and other refinements that give relief effects on walls.

Bricks manage time beautifully. They can last nearly forever. Their rough surface takes a handsome patina that keeps improving for centuries. Walls of brick invite and then record

alterations. You can read in the bricks that an original arched window was later bricked in, then was partially opened and extended to a wide doorway with a stone lintel, then was narrowed a bit, and so on. Recycled brick is considered to have aesthetic as well as economic virtue. The mortar-mottled look of reused brick is so valued that Americans often fake it with blotchy white paint on new brick. The British don't do that. But many of the samples at London's brick library offer fake patina or a fake handmade look, and many a medieval English structure prominently displayed recycled Roman brick.

Recent brick walls, however, may be shorter-lived than their ancestors. Since the early 1960s, most brick walls have been built with a cavity of some 2 inches between an outer layer of decorative brick and the inner layer of cheaper brick or concrete block. (The visual giveaway is that the exterior wall is usually all "stretcher bond"—you see only the long side of the bricks since none are extending inward to form a bond with inner bricks.) The short-term advantages are many—notably higher strength for lower cost and better protection against wind-driven rain. Any water that gets through the permeable bricks of the outer layer will encounter the "rainscreen" protection of the cavity and drip down to the bottom of the cavity, there to return to the outside via weep holes.

The vulnerable component of this system is the metal ties that reach across the cavity to hold the two layers together. When they corrode, the walls become much weaker, with no visible indication that there is any problem until parts of the outer layer fall off or the entire wall collapses. Inspection requires hiring professionals, and correction with new ties is horrendously expensive. Having started with cavity walls in the late 19th century, Britain has more experience with the problem than the US. According to an alarming little book, *Cavity Wall Tie Failure*, in the some 12 million cavity-wall brick buildings in England, tie failure will be found in 70 percent of the buildings from the 1800s, 35 percent in the walls built from 1920 to 1950, and a shocking 40 percent in the ones from 1950 to 1981.[20] Since 1981 more durable

19   Quoted from a study by William H. Ducker in "Masonry Gaining in Nonresidential Work," *Masonry Construction* (Jan. 1991), p. 29. The next most popular cladding materials were "pre-engineered profiled metal," at 24 percent, and "exterior insulation finish systems" at 15 percent and growing rapidly—these are the synthetic stucco-coated foam panels such as Dryvit.

20   Malcolm Hollis, *Cavity Wall Tie Failure* (London: Estates Gazette, 1990), p. 18.

James E. Perkins. Published in his *One Man's World: Popham Beach, Maine* (Freeport ME: Bond Wheelwright, 1974).

ca. 1905 - Popham Beach was a popular spot at an exposed bay mouth on the Atlantic coast of Maine. Attracting the leisure class in the summer called for buildings with romantic fantasy appendages and ornament profoundly unsuited to the Maine winters. Pretty soon: "Does anybody ever go up in that thing? It's falling apart and it's leaking like crazy. Let's get rid of it."

A HIGH-MAINTENANCE BUILDING. Exterior wood bricabrac such as this Italianate belvedere and ornate porch is a pain to keep painted and an extravagance to try to replace when the rot gets to it. The story of the Perkins-Stacey-Spinney House (1885) in Popham Beach, Maine, is the story of a series of owners each finding a new level of surrender—giving up on the belvedere, giving up on part of the porch and filling in the rest, trying both asbestos and aluminum siding, putting in metal-framed windows...

ties have been required, but the expected design life is still on the order of 50 years. Stainless steel ties, with an effective life yet to be determined, are by far the best. Nevertheless, brick cavity walls now share the vilest attribute of aluminum and vinyl siding: they hide their problems.

Concrete is a miracle material, almost. An early ad for Portland cement trumpeted the advantages: "Fireproof, Watertight, Durable, Verminproof, Sanitary, Weatherproof, Rigid, Rapidly Built, No Repairs, No Painting." They left out: Cheap and

6 May 1991. Brand.

1991 - The impact of Atlantic storms and constant erosion of tourist traffic wore the building down to a synthetic nub of its old self. It headed downscale and now provides low-cost overnight lodging. A once-gay building became depressing just because it demanded too much maintenance.

Versatile. And they left out: Impossible To Change. Concrete is so solid and so laced with steel reinforcing that cutting a new door or window is virtually unthinkable. And people don't much like being next to bare concrete. It reeks of prison. But maybe one day, as Brian Eno suggests, "Stained concrete and dirty steel will look rather quaint and friendly and welcoming, like exposed brick does now." The material will certainly get its chance to become popular. According to one enthusiast, "Concrete is second only to water as the world's most heavily consumed substance. Slightly more than a ton of concrete is produced every year for each human being on the planet, some six billion tons a year altogether."[21]

Is concrete really zero-maintenance, as advertised? Like all masonry, concrete is subject to deterioration problems such as (alphabetically): blistering, chipping, coving, cracking, crazing, crumbling, delamination, detachment, efflorescence, erosion, exfoliation, flaking, friability, peeling, pitting, rising damp, salt

---

[21] John Sedgwick, "Strong But Sensitive," *Atlantic* (April 1991), p. 70.

1992 - An arched bridge across Tassajarra Creek leads to the new bath house containing the plunges and baths that are a major attraction for summer guests and a great comfort for the resident Zen students in the winter. When the previous fifty-year-old bath house became dangerously rotted, Berkeley architect Mui Ho was hired to design the new one, after she prevailed in a competition.

1992 - Even though the beams and rafters are made of Port Orford cedar, which is prized for its rot-resistance, checking became a growing problem. In 1990, three years after construction, the absorbent beams ends were painted white. The fix was accompanied with the usual maintenance regret: "We only wish we had done it sooner."

fretting, spalling, subflorescence, sugaring, surface crust, and weathering—most of these caused, as usual, by water.[22] Unlike brick buildings, concrete structures are seldom kept up or repaired. Once they become decrepit, ugly, or irrelevant, they are either demolished with vast noise and expense or left to become particularly unattractive ruins. Concrete is treated like nuclear power: we try not to think about decommissioning.

Roofs should be as close to no-maintenance as possible, but walls, it appears, are better if they're low-maintenance rather than no-maintenance. While roofs protect us, walls channel our lives perhaps too much. Wood and brick walls invite change by involving us in their upkeep. What begins as repair easily becomes improvement—since the brick is being repointed anyway, let's widen one of the windows and replace the double-sash that always gets stuck with a pair of pretty casements, and what about some shutters?

ELEGANT FIX. The handsome unfinished beams and rafters of the new bath house (1987) at Tassajarra Hot Springs in California started cracking badly because of constant exposure to steam and severe dryness. Since Tassajarra is a Zen monastery, the maintenance guy, Stanley Dudek, was familiar with the practice at Japanese temples of sealing beam and rafter ends with paint or metal to keep out moisture that leads to cracking and rot. So he painted the bath house beam ends white. It fixed the problem and added a bright grace note to the look of the building.

An architect who has thought a great deal about the role of maintenance in building evolution is Berkeley's Chris Alexander. The following mini-lecture is pieced together from a conversation we had one afternoon in his backyard.

"Buildings should be really long-lived. The foundation and main frame of the building ought to be built of solid stuff that is capable of lasting three hundred years. Because if the backbone of the building isn't like that, it's very unlikely that you can do all this tampering with it and still leave it in a whole state.

"The kind of building that comes to mind as being the most successful is an Italian or Greek building built out of stone rubble with plastered surfaces. You're looking at a building that is easily capable of being a thousand years old and absolutely serviceable. The core of the building is rather crude, but the finishes are quite highly finished and very easy to put on again. It's possible to make changes at an extremely high level of perfection, even though it might be an informal building.

"It's one thing with a building if it's going downhill and you know you can fix it and bring it into a perfect state, but if it's *generally* starting to go downhill, and you know that it's not going to last more than thirty years, then the motivation is not there.

"On one level, stud construction is very ingenious, very cheap, and it is easy to change, but the building wears down so fast. I've started trying to build with heavy stud construction. I'm building one at the moment out of 2-by-10s and another out of 4-by-6s, just with the idea that solidifying stud construction a bit more may get a better result. But it's still not solid enough.

"There are some very deep interactions with the mortgage system. In order for this kind of upgrading to happen, you have to maintain a constant flow of money into the building. If you are paying on a mortgage, right at the beginning you've skimmed off all your available resources.

"When we did the master plan for the University of Oregon campus, we said that the money should be spread so that both big and small projects would go on over time, half for a few big projects, half for many small projects. That was too tough for people to swallow. They said, 'Fascinating.' It wasn't implemented. The basic idea is that the campus will be in a good condition when it has not only big building projects that are gradually adding to it, but also a continuous series of adaptations—small, very small, and tiny, in ever larger quantities—so that by the time you get down to the smallest level, you've got hundreds of things that are getting tuned all the time. A bench here, a window here, a tree here, a couple of paving stones here. Under those conditions you could say, 'This thing is actually going to get healthy.' If that small stuff isn't happening all the time, you're not going to take care of it, and it isn't going to come to order."

In his book *The Oregon Experiment*, Alexander elaborated: "Large-lump development is based on the idea of *replacement*. Piecemeal growth is based on the idea of *repair*. Since replacement means consumption of resources, while repair means conservation of resources, it is easy to see that piecemeal growth is the sounder of the two from an ecological point of view. But there are even more practical differences. Large-lump development is based on the fallacy that it is possible to build perfect buildings. Piecemeal growth is based on the healthier and more realistic view that mistakes are inevitable.... Unless money is available for repairing these mistakes, every building, once built, is condemned to be, to some extent, unworkable.... Piecemeal growth is based on the assumption that adaptation between buildings and their users is necessarily a slow and continuous business which cannot, under any circumstances, be achieved in a single leap."[23]

Maintenance, in this light, *is* learning.

The three things that change a building most are markets, money, and water. If you would ensure a building's longevity, protect it from markets and water, and feed it money, but not too much and not too little. Too much encourages orgies of radical remodeling

[22] National Park Service, *A Glossary of Historic Masonry Deterioration Problems and Preservation Treatment* (Washington DC: Government Printing Office).

[23] Christopher Alexander, et al., *The Oregon Experiment* (New York: Oxford University, 1975), pp. 77-79. See Recommended Bibliography.

that blow a building's continuity and integrity. Too little, and a building becomes destructive to itself and the people in it. Pressure to cut a building's running costs inspires such shortcuts as cheaper air filters, replaced less frequently, and a lower rate of air cycling, and mildew left in the ducts. Then suddenly some month everybody has headaches and allergies, people are taping ice cubes to the thermostats, and the union's lawyers are on the phone. The amount that should be budgeted annually for maintenance in a commercial or institutional building typically ranges from $2 to $5 per square foot per year, says Chuck Charlton, one of the newly emerging profession of "facilities managers."

They used to be janitors and building superintendents. They were the bottom of the status hierarchy around buildings, but they retained stolid, jealous power. From their private warrens down by the furnace room in the basement of commercial and institutional buildings radiated waves of truculence and an open avoidance of maintenance and repair chores. All that changed in the 1970s with the coming of the information economy— offices multiplying everywhere and computers driving the offices. "Facilities managers were created by information technology," says Frank Duffy, who honors them more than his fellow architects. "An office building exists to accommodate changing organizations. The management of that change process is now the domain of the facilities managers."

By 1979 there was an International Facilities Management Association. It listed nine areas of responsibility which sound dry, but they are the essence of the ongoing life of a working building—planning and design; construction and renovation; coordination of facility changes and relocation; purchasing furnishings, equipment, and external services; developing facilities policies; long-term planning and analysis; building operations, maintenance and engineering; furnishing and equipment inventory management; and real estate procurement and disposal. [24] Unlike the janitors of old who would never change

the filter of the main fan in the building and didn't even possess a manual for the fan, facilities managers typically maintain a log of all filter-requiring equipment in the building, and they hew to a strict filter-replacement schedule.

Facilities managers can be a source of relief or of maddening frustration for building users, depending on how the organization works. If you have to send a minor repair or change request up through channels, wait five weeks, and then get something inappropriate that costs three times what you could have done yourself, the system is not working. In the best operations, still rare, any responsible person in the building can send a work request direct to the maintenance people and get a quick response, and the facilities manager reports directly to the president. Buildings in general should imitate the practice of factories, where the building itself is considered to be the company's most basic and expensive tool, and it is treated with respect and close attention as a profit center. Ideas from the shop floor are sought and heeded.

In factories, for example, the "as-builts" are scrupulously updated. As-builts are building plans that show in detail exactly what *was* built, which is always significantly different from what was in the original plans. Without accurate as-builts, says Chuck Charlton, "An electrical failure can have you wandering through the building shotgunning circuit breakers and shinnying down the chases." Most as-builts that he encounters are at least 10 percent wrong, which multiplies maintenance and remodeling expenses. If the as-builts aren't updated constantly, each bit of repair or remodeling, each new contractor, each change of property management makes the plans more misleading.

This kind of close, sustained attention to the cumulative effect of sporadic bursts of maintenance and repair is essential to conscious "learning" in a building. It ensures continuity of technical processes (layers of paint don't repel each other) and a growing body of lore about what the building needs. A good maintenance log should document who did the work (and their phone number) and precisely what materials were used and where they came from, right down to the brand, color, and supplier of the paint. Each piece of working equipment should have its manual stored carefully either next to it or in a central archive.

Computers are ideal for keeping dynamic records of this sort. Facilities managers are beginning to get sophisticated software that combines the benefits of CAD (computer-aided design), as-built plans, and databases to make a living electronic model of the building in finest detail:

> After a roof is surveyed, graphic and nongraphic information is stored showing the age of the roof and a projected replacement date. By querying the system about a certain part of the roof, we can review tabular information including roof type, R-value [insulation factor], insulation type, original installation date, warranty date, contractor, and consultant. This data improves planning, budgeting, and reporting.[25]

It is easy to update, so even minor changes get entered. Homeowners should have such tools.

Though I personally favor old or Low-Road buildings whose systems and materials are intuitively obvious, the fact is that most buildings are increasingly complex and high-tech. As they become more complicated, their subsystems require the attention of specialists to inspect and maintain. The researchers of *The Occupier's View* noted, "We noticed that the number of problems often seem to increase in direct proportion to the level of sophistication of the [air and heat] system. This was expressed by one company as follows, 'The more complicated the system the greater the chance of things going wrong.'"[26]

Even in the home do-it-yourself can do a lot, but not everything. I recall a systems story told with telegraphic compression on a computer teleconference system:

> A story out of Houston a number of years ago. People in a house didn't pay the gas or electricity. Gas company shut off the gas. People turned it back on. Gas company came out,

removed the meter. People took radiator hose, ran it between gas line and house line. They were using candles for light, since the electricity was shut off. Gas meter contains an important device: pressure regulator. Gas in pipeline is at approximately ten times the pressure after the meter. Regulators in gas stove couldn't take it, blew off. Gas mix found candles. I disremember the fate of the people. House vanished.[27]

There is a sensing problem with buildings. Too much is invisible—the pressure regulator in the gas meter, the rot in the walls, the location of the short circuit. Ventilation is especially elusive. While people are acutely sensitive to temperature problems and always ready to bang on a thermostat, they don't notice when they aren't getting their requisite fifteen cubic feet per minute of fresh air. Manufacturers eager to sell electronic products are beginning to produce a variety of systems to sense and report incipient maintenance problems. Buildings will become automatically self-diagnosing like an office copier or an airplane, and that's fine. But I'd like to see building designers take on *problem transparency* as a design goal. Use materials that smell bad when they get wet.[28] Build in inspection windows and hatches. Expose the parts of service systems that are likeliest to fail.

The all-time best models of superb maintenance, both in terms of design and administration, are hospitals and naval ships. For them maintenance issues are "life-critical," and it shows. Inspection is routine and exhaustive. Maintenance tasks are

---

[24] Steelcase Strafor, *The Responsive Office* (Streately-on-Thames, Berkshire: Polymath, 1990), p. 66.

[25] B. J. Novitski, "Facility Management Software," *Architecture* (June 1991), p. 114.

[26] *The Occupier's View*, p. 122.

[27] Martha Elaine Sweeney, Homeowners Conference on The WELL, May 1989.

[28] An example of making failure conspicuous is the practice of natural gas suppliers, who scent their invisible, odorless, dangerous gas with the sharp stench of ethyl mercaptan, so people will detect leaks and be motivated to fix them.

divided into "preventive" and "corrective." In-house maintenance staffs are on call 24 hours a day, and all working equipment is kept in good-as-new condition. So far as I know, ships and hospitals have never been studied as design exemplars for other kinds of buildings.

Too often a new building is a teacher of bad maintenance habits. After the initial shakedown period, everything pretty much works, and the owner and inhabitants gratefully stop paying attention to the place. Once attention is deferred, deferring of maintenance comes naturally. It might be better if some of the original work were intentionally ephemeral, with everyone knowing it will require maintenance or replacement within a year. "How might a new building teach good maintenance habits?" is a question worth giving to architecture students.

Maintenance comes in two major flavors, especially around houses—cosmetic and real. Unfortunately the cosmetic is more fun. It's like the weekend sailor who puts loving attention into his sailboat's varnished brightwork and lets the engine rust. Serious sailors paint over the brightwork and lavish their fretful attention on the engine, laying in a spare water pump and extra belts. Maybe the trick for homeowners is to mix serious and frivolous chores: replace the air filter in the furnace, then go putter in the garden. Or be sure that any repair includes the reward of some improvement. The temptation to avoid is concealing the need for real work with a cosmetic touch-up—painting the rot.

Deborah Devonshire accords high status to the keepers of Chatsworth and its lands and celebrates their tasks:

> In the house and out of doors vigilance and maintenance, unseen and unsung, are the order of the day's work. Nothing is permanent. Lead on the roof wears thin, and a hole the size of a pinhead lets in rain which can soon turn into dry rot. Stone, especially when bedded the wrong

way of its grain, flakes, and the weather finds the weak places and scoops them out as if with a giant spoon.... Wormwood, death-watch beetle, fire, water, snow, frost, wind and sun (All Ye Works of the Lord, in fact) each does its special harm.[29]

Against the flow of this constant entropy, maintenance people must swim always upstream, progressless against the current like a watchful trout. The only satisfaction they can get from their work is to do it well. The measure of success in their labors is that the result is invisible, unnoticed. Thanks to them, everything is the same as it ever was.

The romance of maintenance is that it has none. Its joys are quiet ones. There is a certain high calling in the steady tending to a ship, to a garden, to a building. One is participating physically in a deep, long life.

The anthropologist/philosopher Gregory Bateson used to tell a story:

> New College, Oxford, is of rather late foundation, hence the name. It was founded around the late 14th century. It has, like other colleges, a great dining hall with big oak beams across the top, yes? These might be two feet square, forty-five feet long.
>
> A century ago, so I am told, some busy entomologist went up into the roof of the dining hall with a penknife and poked at the beams and found that they were full of beetles. This was reported to the College Council, who met in some dismay, because where would they get beams of that caliber nowadays?
>
> One of the Junior Fellows stuck his neck out and suggested that there might be on College lands some oak. These colleges are endowed with pieces of land scattered across the country. So they called in the College Forester, who of course had not been near the college itself for some years, and asked him about oaks.

29 Deborah Devonshire, *The House* (London: Macmillan, 1982), p. 83.

And he pulled his forelock and said, "Well sirs, we was wonderin' when you'd be askin'."

Upon further inquiry it was discovered that when the College was founded, a grove of oaks had been planted to replace the beams in the dining hall when they became beetly, because oak beams always become beetly in the end. This plan had been passed down from one Forester to the next for five hundred years. "You don't cut them oaks. Them's for the College Hall."

A nice story. That's the way to run a culture.

Every time I've retold this story since I first heard it from Gregory in the 1970s, someone always asks, "What about for the next time? Has a new grove of oaks been planted and protected?" I forwarded the question to the authorities at New College—the College Archivist and the Clerk of Works. They had no idea.

September 1990. Brand.

1990 - The new oak beams of New College, Oxford, were installed in 1865 by Gothic Revivalist Gilbert Scott, using timbers from college estates at Great Horwood and Akely in north Buckinghamshire. The building is the oldest surviving college hall at Oxford, completed in 1386 by Bishop William of Wykeham (master mason, William Wynford). The now-windowed opening in the roof was originally to let out smoke from an open fire in the center of the hall.

CHAPTER 9

Vernacular: How Buildings Learn From Each Other

WHAT GETS PASSED from building to building via builders and users is informal and casual and astute. At least it is when the surrounding culture is coherent enough to embrace generations of experience.

"Vernacular" is a term borrowed since the 1850s by architectural historians from linguists, who used it to mean "the native language of a region." Chris Alexander adopts a similar usage when he declares that a "pattern language" is the medium of humane building design. "Vernacular" means "vulgar" sometimes and "the bearer of folk wisdom" sometimes. It means "common" in all three senses of the word—"widespread," "ordinary," and "beneath notice."

In terms of architecture, vernacular buildings are seen as the opposite of whatever is "academic," "high style," "polite." Vernacular is everything not designed by professional architects—in other words, most of the world's buildings, ranging in assigned value from now-precious Cotswold stone cottages and treasured old Cape Cods to the despised hordes of factory-built mobile homes. In the eyes of tastemakers, old vernacular is lovely. New vernacular (including everything we might call Low Road) is unlovely.

There is a magazine called *Progressive Architecture* but none called *Conservative Architecture*. If there were such a magazine (a good idea, in my view), it would be largely about vernacular architecture, which is profoundly cautious and imitative, so immersed in its culture and its region that it looks interesting only to outsiders.

Eminent folklorist Henry Glassie has described vernacular

building tradition in operation:

> A man wants a house. He talks with a builder. Together they design the house out of their shared experience, their culture of what a house should be. There is no need for formal plans. Students of vernacular architecture search for plans, wish for plans, but should not be surprised that they find none. The existence of plans on paper is an indicator of cultural weakening. The amount of detail in a plan is an exact measure of the degree of cultural disharmony; the more minimal the plan, the more completely the architectural idea abides in the separate minds of architect and client.[1]

Vernacular building traditions have the attention span to incorporate generational knowledge about long-term problems such as maintaining and growing a building over time. High-style architecture likes to solve old problems in new ways, which is a formula for disaster, according to Dell Upton at the University of California. Vernacular builders, he says, are content to accept well-proven old solutions to old problems. Then they can concentrate all their design ingenuity strictly on new problems, if any. When the standard local roof design works pretty well, and materials and skills are readily available for later repair, why would you mess with that?

Vernacular buildings evolve. As generations of new buildings imitate the best of mature buildings, they increase in sophistication while retaining simplicity. They become finely attuned to the local weather and local society. A much-quoted dictum of Henry Glassie's states that "a search for pattern in folk material yields regions, where a search for pattern in popular material yields periods."[2] Roof lines and room layout are regional.

ca. 300 BC - Reconstructed from archaeology at Groningen in The Netherlands, this typical combination of house and barn shows how the wide center aisle (nave) is used for circulation and common uses such as the fire, while the side aisles can be subdivided for specialized or private functions.

Walter Schwarz, 1958. Reprinted from The Plan of St. Gall, vol. 2, p. 44.

ca. 820 AD - Reconstructed from detailed plans for a Benedictine monastery at St. Gall, Switzerland. As depicted in the plan for this House for Distinguished Guests, servants occupied the left aisle around the side entrance, while horses were stabled in the aisle flanking an exit to privies on the right. Guests had private rooms at each end and dined in the common area around the open fire in the nave.

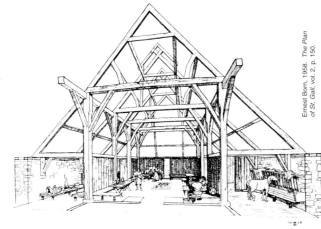

Ernest Born, 1958. The Plan of St. Gall, vol. 2, p. 150.

ca. 1295 AD - St. Mary's Hospital in Chichester, England, still survives (see next page). The six-bayed infirmary hall opened into a chapel at the far end. A hospital at that time provided shelter for pilgrims and paupers as well as the ill. No medical treatment was attempted.

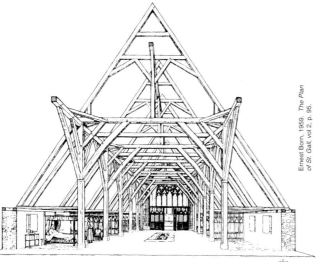

Ernest Born, 1959. The Plan of St. Gall, vol. 2, p. 95.

THREE-AISLED STRUCTURES in northern Europe date at least from 1300 BC and continue their evolution to this day. They were robustly multi-purpose—Nordic sagas chronicle their use as dwellings, dining halls, celebration halls, kitchens, dormitories, and barns for cattle or hay. Often many of the uses combined in one building or followed each other in sequence. These drawings are from the impeccable work, *The Plan of St. Gall*, by Walter Horn and Ernest Born (see Recommended Bibliography).

Paint color and trim vary with fashions in style. The heart of vernacular design is about form, not style. Style is time's fool. Form is time's student.

You can see it in two major forms of European rural building of ancient lineage. The Mediterranean masonry courtyard house has long been celebrated for its perfect fit with local climate and landscape and the multi-generational needs of extended families. Less well-known but equally admirable is the long, three-aisled, steep-roofed wood structure developed in the forests of the north. Here the courtyard equivalent—the double-wide aisle in the middle with its central fire—was enclosed from the inclement weather. With the thatch-bearing rafters supported at their center of load by vertical timber columns, the buildings were easy to raise and then easy to adapt. The side aisles invited subdivision for a variety of purposes and for privacy. Growth came easily by adding bays at either end, and yet the interior was always kept unified by the wide central aisle. The form has survived the

---

[1] Henry Glassie, "Vernacular Architecture and Society," *Vernacular Architecture: Ethnoscapes: Vol. 4*, Mete Turan, ed., p. 274. This is exactly the point that Christopher Alexander makes in *The Timeless Way* and *A Pattern Language*.

[2] Henry Glassie, *Pattern in the Material Folk Culture of the Eastern United States* (Philadelphia: Univ. of Pennsylvania, 1968), p. 33.

1991 - The tithe barn at Great Coxwell, Oxfordshire, is 152 feet long, 43 feet wide, 48 feet high. William Morris, who lived nearby, regarded it as "the greatest piece of architecture in England." The sophisticated bracing of beams and roof evident here is thought to have worked out in wood the forms later used in stone cathedrals.

9 August 1991. Brand.

ca. 1910 - St. Mary's Hospital (see previous page) was converted to an almshouse in 1535, offering private rooms. These partitions and chimneys date from 1680. In the four bays remaining (of the original six) are eight two-room dwellings still used for old people, and the center aisle still opens into the chapel.

Royal Commission on the Historic Monuments of England. Neg. no. BB 89/9520.

TIMBER-FRAMED ROOFS, steep-pitched, are perfectly capable of lasting seven centuries, as in the Great Coxwell barn (ca. 1310) and St. Mary's Hospital (ca. 1295). Their three-aisled form kept them useful.

millennia magnificently in barns and cathedrals and is worth reviving for houses.

The lesson for the ages from three-aisled structures is that columns articulate space in a way that makes people feel comfortable making and remaking walls and rooms anchored to the columns. You can always visualize what you might do next to improve the space plan. The recent engineering triumph of huge free-span interior spaces is actually a loss for intuitive adaptivity. The effect of wide-open space is oppressive rather than freeing.

The space plans of vernacular buildings are typically generic and general-purpose. The identical bays of three-aisled structures and the additive identical rooms of courtyard houses had been found to be the most inexpensively adaptable over time. Vernacular design is always prudent about materials and time, seeking the most pragmatic building for the least effort and cost. It provides an economical grammar of construction. Let there be a central passageway and stair hall, say, with roughly identical pairs of rooms on each side upstairs and down. (That was the "double pile" house that the fathers of George Washington and James Madison built and that pervaded eastern America.) The specifics of material, style, and finish were left to the builder and dweller.

To the cultural historian Ivan Illich, the spare clarity of such

3 Ivan Illich, *In the Mirror of the Past* (London, New York: Marion Boyars, 1992), p. 56.

4 Dell Upton introducing Thomas Hubka, "Just Folks Designing," *Common Places*, Dell Upton, John Michael Vlach, eds. (Austin, GA: Univ. of Georgia, 1986), p. 426.

5 Title above, pp. 431 and 433.

buildings was honed by countless real and individual lives:

> Dwelling is an activity that lies beyond the reach of the architect not only because it is a popular art; not only because it goes on and on in waves that escape his control; not only because it is of a tender complexity outside of the horizon of mere biologists and system analysts; but above all because no two communities dwell alike. Habit and habitat say almost the same. Each vernacular architecture…is as unique as vernacular speech. The art of living in its entirety—that is, the art of loving and dreaming, of suffering and dying—makes each lifestyle unique. And therefore this art is much too complex to be taught by the methods of a Comenius or Pestalozzi, by a schoolmaster or by TV. It is an art which can only be picked up. Each person becomes a vernacular builder and a vernacular speaker by growing up, by moving from one initiation to the next in becoming either a male or a female inhabitant. Therefore the Cartesian, three-dimensional, homogeneous space into which the architect builds, and the vernacular space which dwelling brings into existence, constitute different classes of space.[3]

And they were arrived at by different classes of design. The process of vernacular design is treated, even by its admirers, with undeserved condescension, insists building historian Thomas Hubka. In an introduction to Hubka's paper "Just Folks Designing," Dell Upton summarizes:

> Hubka carefully distinguishes the vernacular builder's process of design, in which existing models are conceptually taken apart and then reassembled in new buildings, from the professional designer's manner of working, in which elements from disparate sources are combined to solve design problems anew. He characterizes the vernacular architect's process as "preconstrained"; by *choosing* to limit architectural ideas to what is available in the local context, the vernacular architect reduces the design task to manageable proportions.

Although this mode of composition seems superficially to generate monotonously similar structures, it allows in fact for considerable individuality within its boundaries, permitting the designer to focus on skillful solution of particular problems rather than on reinventing whole forms.[4]

Hubka says in the article that, far from constricting the folk builder's own creativity and individuality, this approach frees them:

> The folk designer simply signs his signature much smaller [than contemporary designers] but by no means less forcefully. This signature is in the details, in the care, and in the craft of building (and while the modern observer might not see this signature you can be sure his contemporaries saw it). Folk architecture that appears unified, homogeneous, even identical becomes, on closer inspection, rich, diversified, and individualistic.

Hubka concludes: "A case can, and should, be made for folk design method as one of the most pervasive and well-conceived design methods in the history of civilization."[5] He recommends that contemporary architects study it.

I was drawn to talking with vernacular building historians because, more than other architectural historians, they focus on how the buildings *work*. In America the discipline is new and young—some call it a "children's crusade"—having been pioneered in mid-20th-century by the likes of J. B. Jackson (vernacular landscapes) and Henry Glassie (folk material culture) and fostered by the growth of the preservation and environmental movements.

Vernacular building historians excel at "reading" buildings—analyzing the physical evidence of what actually happened in a building, and when, and why. I inquired about the tricks of the trade with Orlando Ridout V, head of the Office of Research, Survey and Registration of the Maryland Historical Trust, based in

136

ca. 1885 - Pump Square of Siasconsett, Nantucket, founded by whalers in the 1680s. The full taxonomy of add-ons is displayed. A former shanty in the foreground has become an ice cream saloon.

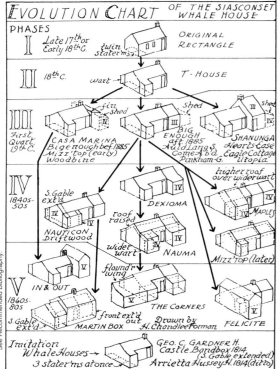

Henry Chandlee Forman's chart of how the whale houses grew (viewed as somewhat fanciful by other building historians). The original "great room" of these houses was only 11 by 13 feet.

INCREMENTAL GROWTH. Because vernacular houses assume the inevitability of later expansion and always seek the economical path, they are universally expert at growing by stages. The semi-medieval "whale houses" of the island of Nantucket off Massachusetts were so tiny they had to grow, and they grew in a locally-patterned way. The practice reached its celebrated apogee in the big-house-little-house-back-house-barn "connected farms" of the mid-19th century.[6]

1940 - The "connected farms" of mid-19th-century New England rationalized add-ons via a popular theory of more efficient agriculture. Like most, this one in Monticello, Maine, cups around a south-facing work yard.

ca. 1985 - Steel-roofed against the tropical rains, raised on stilts for privacy, for breezes, for protection from floods and animals, and for storage underneath, the vernacular house of Malaysia is supremely adapted to its climate and culture.

ca. 1985 - The interior of a Malay house is designed for natural ventilation while excluding glare and rain. Windows are wide and low, roofs have long overhangs, and interior space is wide open. Rooms are indicated by differing floor and ceiling levels, which also makes growing the building by increments exceptionally easy, since floors and roofs don't have to match precisely.

Reprinted from Lim Jee Yuan, *The Malay House,* pp. 147 and 72.

Far more sophisticated, probably because the culture is more stable, are the traditional village houses of Malaysia. The Malay house, with its refined, varied means of growing from the original core house, is a wonder of incremental architecture. Lim Jee Yuan rightly claims, "It created near-perfect solutions to the control of climate, multifunctional use of space, flexibility in design and a sophisticated prefabricated system which can extend the house with the growing needs of the family."[7]

The Malay house has a specific pattern language of growth, with special terms for the core house (*rumah ibu*), same-level verandah (*serambi samanaik*), step-down verandah (*serambi gantung*), covered walkway (*selang*), kitchen (*dapur*), front extension (*lepau*), back extension "like a baby elephant suckling its mother" (*gajah meyusu*), and entrance porch (*anjung*).

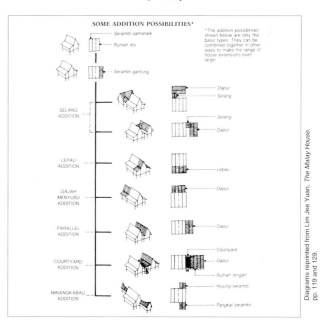

SOME ADDITION POSSIBILITIES

Diagrams reprinted from Lim Jee Yuan, *The Malay House,* pp. 119 and 129.

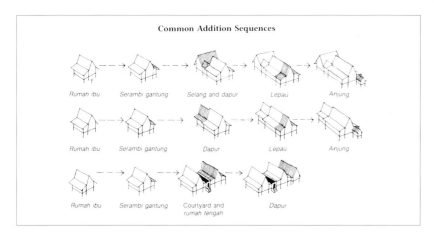

Common Addition Sequences

---

[6] A full exploration of the whale houses of Siasconset is given in Henry Chandlee Forman, *Early Nantucket and Its Whale Houses* (Nantucket: Mill Hill, 1966). See Recommended Bibliography.

The story of New England connected farms and the commercial theory that shaped them is told with style and authority in Thomas Hubka, *Big House, Little House, Back House, Barn* (Hanover: Univ. Press of New England, 1984). See Recommended Bibliography.

[7] The best book that I've seen on any indigenous architecture is Lim Jee Yuan, *The Malay House* (Pulau Pinang: Institut Masyarakat, 1987). See Recommended Bibliography.

BUILDING HISTORIAN Orlando Ridout points out what skilled masons did for a 1774 triple townhouse in Annapolis, Maryland, designed by his ancestor John Ridout: "When they struck the joints, see how they pulled the mortar away from the brick, and then they added this little incised line in the center of the joint, and it gives a very precise, neat look. When the sun is coming down on it, it casts a shadow and it punches the wall really nicely."

RIDOUT: "We're looking at number 80 and 82 East Street. We have a house that's gone through one major period of change. There's some indication of that change in the brickwork on the left hand gable end, but the upper gable and entire facade is circa-1840s brickwork. We can see those familiar mid-19th-century wooden lintels over the windows and door, and it has a corbeled brick cornice, which is very typical of the second quarter of the 19th century. It has the plainer, uniform-looking brickwork that we associate with the first half of the 19th century. But we do have some clues that there's something earlier here—changes in the brickwork at the end. We also have the water table breaking out, creating the sense of a real foundation. The stuff in the back was probably added in the late 19th century. It has typical 'German' siding—some people call it 'novelty siding'—from usually the late 19th and early 20th century."

historic Annapolis. A second generation architectural historian and onetime builder himself, Ridout reminded me of field biologists and geologists. For such people the world is a constant puzzle and revelation, filled with what Vladimir Nabokov called "transparent things, through which the past shines!"

As we strolled around Annapolis, Ridout explained how to see through a building. "Traditionally architectural history, because it came from art history, has tended to focus on style. Style is the last thing that I would teach a student about architectural history, because it's so misleading. I could care less what style a building is. I want to know when it was built, and how it evolved, and what floor plan it had, and how the spaces in that house were used. The best way to approach dating a building and unraveling the sequence of change is to look at things that are least likely to lie to you—essentially, the things that are least self-conscious. The living room is a self-conscious part of the house. The front facade is a self-conscious part of the house, where the owner is trying to make a statement to the world about what he is about—whether it's 'I'm a simple man with simple tastes' or 'I'm richer than you are and don't you forget it' or simply, 'I have crossed the threshold of gentility. I can now afford a brick house with a fancy entrance porch.'

"So you head for the parts of the house where nobody has tried to dress things up, and that's the attic and the cellar. In the attic what you're really looking for is where a second period of construction has encapsulated an original period of construction. Even the windows sometimes get buried in place, and the paint colors are all still there. The most original hardware's always at the top of the house. It got the least amount of use, and nobody cares if it doesn't look great, so it stays there. Go up and look at the servants' quarters.

"The thing that is most helpful in figuring out a building's history

9 May 1991. Brand.

"If we look at the right hand gable end of the building, where the stucco has been stripped away, we can see very clearly, part way up, the 18th century gable end of the building, with evidence of at least three small windows and very nice Flemish bond brickwork, and glazed headers forming a diamond pattern in the center of the wall.

"Stucco can be a real problem for people, because they don't like it aesthetically, and so they take it off, not realizing that usually stucco went on to solve a problem or to cover up messy aesthetics. One reason for stucco is, if you've got water penetration problems, it's the cheapest, fastest fix. In the 1830s it was popular to stucco a building anyway, so often they were built with crummy brick, or broken brick, reused brick. People later spend thousands of dollars to tear all the stucco off, and then they've got a wall that's worthless and they've got to put it all right back up.

"It's a standing seam metal roof. Beginning in about the middle of the 1840s you see quite a bit of it, especially in dense urban neighborhoods because of the fire prevention qualities."

is technology. Building technology never lies. No 18th-century builder ever had access to a machine-made nail. He had to use a handmade nail. He'd never seen a circular saw. He couldn't buy his boards planed in a planing mill or his plaster lathing cut on a bandsaw. An 18th-century house is totally a hand crafted building, with very distinctive tool marks, methods of construction, joinery, nails, interior finishing elements—whether it's the plaster lathing or the trim around the door or the way he framed the base of the chimney. The industrial revolution begins to show up in the late 1790s, but it really doesn't begin to take hold until the 1840s and 1850s in the Chesapeake area. After the Civil War you have a sea change. Then you're basically building national housing."

I asked Ridout what the historic building inventories and archaeology of Maryland suggested about the comparative survivability of the various kinds of old buildings. He said the main survivors were masonry buildings, even though they were only 15 percent of what was originally built. Medium to large houses survived the best, because there is always use for them. Small houses were built shoddy and disposable. Barns survived fairly well, thanks to being solidly constructed. Specialized farm buildings perished of obsolescence, with one interesting exception.

"Small domestic outbuildings that are well built tend to survive. Everybody can find a use for a 12- or 14-foot-square building. Meat houses, for example, tended to be very well built. They were either of masonry, log, or very heavy timber frame construction, partly because they were carrying a lot of weight— maybe 2,000 pounds of ham hanging from the roof—and they had to be theftproof. They usually were relatively close to the main house, so they're still convenient to this day. These days they're full of lawnmowers and weedcutters and turpentine and bicycles. They're very hard to measure because they're always crammed with junk."

Vernacular building historians of the current breed are interested in any kind of building from any period, including the present. The patriarch J. B. Jackson has remarked, "The older I get, the more interested I get in the future that's waiting for us. I don't think it will have much dignity, but it will have vitality."[8] And vitality is what he inspired people to study.

One of the things that would be worth investigating in contemporary buildings is the *informal* pathways of influence. The formal pathways of architect influencing architect have been

---

8 Interview with Jane Holtz Kay reprinted from the *New York Times* in the *San Francisco Chronicle* (21 Sept. 1989). J. B. Jackson's influential books include *The Necessity for Ruins* (Amherst: Univ. of Mass., 1980) and *Discovering the Vernacular Landscape* (New Haven: Yale Univ., 1984).

studied to death, but they explain little about where most of the real action is. Even in matters of style, some elements seem to have lives of their own, like classical columns. A decorative Post-Modern column refers to the Beaux Arts column, which referred to the Renaissance column, which referred to the classical Roman column, which referred to the classical Greek column of stone, which referred to the earlier wooden column made of a tree. They are all mostly nonfunctional and expensive to craft. Clem Labine, founder of *Old House Journal*, has commented, "While the popularity of classicism has certainly waxed and waned, there hasn't been a period in over two millennia when someone in some part of the world hasn't been fitting architraves across column tops."[9] Modernism swore it would get rid of these pagan temple ornaments forever, and the first thing Post-Modernism did was put them back.

Something evidently drives continuity between buildings at a mythic level. Masonry fireplaces and chimneys have been utterly obsolete since the popularization of the Franklin stove by the 1830s, yet 160 years later every house that can afford it still has at least a facsimile of a masonry fireplace and chimney. Some deep lullaby croons, "Hearth and home."

In some high-style buildings the architect decrees and the client accepts—a status battle lampooned in Tom Wolfe's *From Bauhaus to Our House*—but in most buildings it is the other way around. And clients seldom innovate. They borrow. They see something they like, and they insist that their building be "like that." How did running water, bathrooms, central heating, and air conditioning originally get into houses? Not via architects. Historian Daniel Boorstin tells an interesting tale of the dialogue between public and private buildings:

> In the older world [of Europe], the *public* facilities tended to copy the *private*. Inns were shaped like large private residences, town halls were fashioned after the palatial dwellings of rich citizens. But the urban communities which sprang up in the United States in the nineteenth century were bristling with newcomers, while there were still few rich men and, of course, no ancient palaces. Here public buildings and public facilities made their own style, which gradually influenced the way everyone lived.[10]

It was raffish commercial buildings rather than the stately institutional ones that led the way. Grand hotels, which historians consider to be an American invention, introduced gas light, spring mattresses, running water and central heating by mid-19th century. Guests soon took insistence on such luxuries home with them. Once people had experienced air conditioning in movie theaters in the 1930s, they could not bear to live without it at home. Maybe home adoption is the final test of the success of a new building element. That the sealed windows and services-hiding dropped ceilings of offices have not migrated to the home may signify that they are ultimately failures.

Sometimes a building form takes off and becomes so widely popular that its design is assumed to be anonymous—"folk"—when in fact it was some individual's bright idea. America's roadside service stations were largely created and continuously updated, right up to the time of interstate truck stops, by one man.

9 Clem Labine, "Please Pass the Civitas," *Traditional Builder* (Dec. 1990), p. 4.

10 Daniel Boorstin, *The Americans: The Democratic Experience* (New York: Random, 1973), p. 350. He has a whole chapter on "The Palaces of the Public" in *The Americans: The National Experience* (New York: Random, 1965), p. 137.

11 Brendan Gill, *Architecture Digest* (May 1991), p. 27.

12 A counter-argument could be made that the eighteen volumes of *Sweet's General Building and Renovation Catalog File* offer way too many products—21,000 pages of stuff from 2,300 manufacturers in 1992, and growing—and they are always too new for any knowledge to accumulate about whether they work or not. This is the opposite of choice-narrowing vernacular design, which knows a few things very well and clings to them.

13 I am indebted, in my bare-bones account of the generation of Santa Fe style, to discussion with Chris Wilson, a cultural historian at the University of New Mexico, Albuquerque. His forthcoming book, *The Myth of Santa Fe: Tourism, Ethnic Identity, and the Creation of a Modern Regional Tradition* (Albuquerque: Univ. of New Mexico), will be a landmark study of the commercialization of vernacular form. A preview may be found in his paper, "New Mexico in the Tradition of Romantic Reaction," *Pueblo Style and Regional Architecture*, Nicholas Markovich, et al., eds. (New York: Van Nostrand Reinhold, 1990), pp. 175-194.

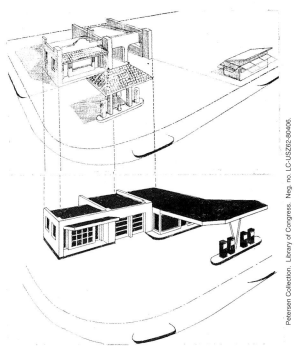

ca. 1945 - CARL PETERSEN not only designed American gas stations for Gulf Oil and Pure Oil from 1914 to the mid-1950s, he designed the radical updating of his own creations. Not an architect, he shaped thousands of buildings that became a vernacular conspicuously characteristic of the nation and the times. When he retired he was designing the first truck plazas for the new interstate highways.

Carl Petersen devised these buildings for several American oil companies from 1914 to 1970, and his designs were copied by the competition. Suburban "ranch houses," which defined the American 1950s, largely came from a little-known Californian named Cliff May. For years his designs wouldn't sell. Then a contractor advised him to stop trying to hide the driveway and garage and instead flaunt them, because Americans love showing off their car. With that his career took off. He personally designed a thousand ranch-style houses and had 18,000 built to his design by developers. "On the strength of these figures, it could be claimed for May that he is probably the most popular architect that has ever lived," observed one critic.[11] Though May never went to architecture school and never got a license, his

work attracted the admiration of Frank Lloyd Wright (a rare event) and the flattery of imitation by countless developers.

It may be that one of the reasons architects are so driven toward surface originality is that their industry compels them to uniformity throughout the rest of a building. Caught between the rigid requirements of building codes, the standard solutions of professional books such as *Architectural Graphic Standards,* and the standard products of *Sweets Catalog,* there is not much room for creativity, so architects grab what they can.[12] This is a gain and a loss for quality in buildings. The worst are less bad because of having to meet fairly intelligent standards. But the best are less innovative as a whole, and they are less likely to be finely adapted, or adaptable, to their unique circumstances. Instead of learning from each other, such "catalog architecture" buildings are guided by a standard homogenized pool of building lore which is no longer regional and often not even national, but world-encompassing, inescapable and unchallengeable.

How else can we explain the survival from decade to decade of the aluminum-frame sliding glass door? It seems to serve simultaneously as door, window, and wall, but it's terrible at all three. As a door it's fiddly and awkward to open, and dangerous, since it has the vicious property of looking the same when open or closed, and people walk smack into the glass. As a window it reveals too much in both directions and makes any view quickly boring. And it's worthless as wall, being nonstructural and noninsulated, bleeding heat in whatever is the wrong direction. The sliding glass door is a measure of how remote the builders' decisions have become from the users' experience and of how powerless users are in the face of standardized building doctrine.

Where a regional tradition does survive into the modern world it can work potent magic, even when adulterated. A textbook case is the triumph in the American southwest of "Santa Fe style" (also called "Pueblo Revival").[13] Visitors today to Santa Fe, New Mexico, are thrilled to find an entire city of low, beautifully sculptured adobe buildings gleaming in the high desert sun,

ca. 1868 - Spanish colonial governor Don Pedro de Peralta had the "*palacio real*" (royal palace) built in the new capital of Santa Fe in 1610 as a rude fort and administrative headquarters. From 1680 to 1692, following the Pueblo Revolt, Tewa and Tano Indians occupied the building, converting it to a multi-story pueblo until the Spanish recaptured it. Americans seized the building from Mexico in 1846 and housed the Territorial Governor there for decades. Sawn lumber was showed off in the *portal* columns.

US Army Signal Corps Collection. The Museum of New Mexico. Neg. no. 9099.

ca. 1882 - Always approaching a ruinous state ever since it was built, the Palace was renovated in 1877 with a metal roof and a fancy balustrade on the *portal*. Decoration on the near end included an elaborate brick cornice and stucco painted to look like stone. The interior was still a mess in 1878 when Governor Lew Wallace moved in. He wrote of his writing room: "The walls were grimy, the undressed boards of the floor rested flat upon the ground; the cedar rafters, rain-stained as those in the dining-hall of Cedric the Saxon, and overweighted by tons and tons of mud composing the roof, had the threatening downward curvature of a shipmate's cutlass. Nevertheless, in that cavernous chamber I wrote the eighth and last book of *Ben-Hur*."

Ben Wittick. The Museum of New Mexico. Neg. no. 15376.

obviously redolent with history, inheritors of America's most ancient building lineage. It is all a 20th-century invention.

Santa Fe style developed from the collision of three vernacular building traditions and one generation of calculating boosters. It was an epic of cross-cultural borrowing. The mythic baseline comes from the Indian multi-story adobe and stone "apartment houses" of New Mexico and Arizona that were built dense and high by the agricultural Pueblo tribes. They were stepped down, terrace by terrace, toward the south and southeast to soak up the sun's warmth. Spanish explorers arrived in the area in 1540 and began colonizing in 1598 (twenty-two years before the Mayflower Pilgrims) with an architecture somewhat similar to the Indian pueblos based on the traditional Mediterranean courtyard house—masonry, flat-roofed, with small general-purpose rooms added casually.

The Indians soon adopted several Spanish innovations. They replaced puddled adobe with wood-formed adobe bricks laced with straw. They replaced smoky open fires in the pueblo rooms

SANTA FE STYLE. America's oldest public building, the adobe Palace of the Governors (1610) on the Santa Fe plaza, was the first expression of the city's determination to redesign itself as a tourist town. One of the devisers of the architectural Indian-Spanish-Anglo amalgam called Santa Fe style was archaeologist Jesse Nusbaum. From 1909 to 1913 he remodeled the *portal* of the Palace backward past its Victorian and Territorial periods to an imaginary colonial look. The result was one of the four or five New Mexico buildings most influential as a model for thousands of Santa Fe style buildings.

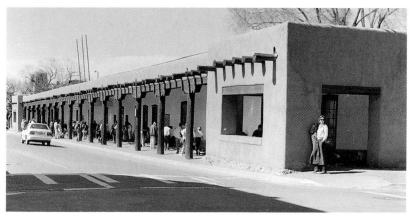

1913 - In 1909 the Palace was turned over to the brand new Museum of New Mexico, and its first employee, Jesse Nusbaum, set about rebuilding the *portal*. In one of the old adobe walls he found embedded a round wood column and corbel, which he took as a pattern for the new columns. Like most Santa Fe style buildings, the Palace *portal* only looks like adobe. As shown in this photo by Nusbaum, he built for the ages with stuccoed brick.

1991 - For nearly a century now the remodeled Palace of the Governors has been one of the major attractions in Santa Fe, with its excellent museum inside and its daily market of jewelry offered by local Indians on colorful blankets under the *portal*. Tradition is what you make it. That is, most traditions were once someone's bright idea which was successful enough to persist long enough for people to forget that it was once someone's bright idea.

with the shapely adobe corner fireplace called a *fogon*, and they began whitewashing interior walls. Outside, they added the Spanish beehive oven, the *horno*, beloved of tourist photographers. Meanwhile the Spanish, imitating the Indians, often opened their courtyard houses into L- and U-shapes facing south and southeast. The colonial capital in Santa Fe was so isolated that not much further influence came from Europe. The nearest Spanish city was thirty days' travel to the south, and trade caravans only came once every three years. Two centuries went by.

The third force, Yankees, started arriving on the Santa Fe Trail after 1821 and brought with them a frontier vernacular and an accelerated pace of change. By 1850 local sawmills were making milled lumber, doors, and windows. Soon came glass and manufactured metal hardware. After 1879 the Atchison Topeka &

Santa Fe railroad brought in even more goods and Anglos (the regional name for non-Spanish whites). They imported the region's first architectural *style*—a derivative of Greek Revival known today as the Territorial style, featuring rudimentary ornamentation such as pediments over doorways and decorative brick atop adobe walls.

While the Anglos were adopting some Spanish practices such as *portales*—covered walkways in front of commercial buildings— the traditional Spanish buildings were becoming thoroughly hybridized. They remained one-story, small-roomed, low to the ground, and casually additive, but they began to acquire pitched roofs, porches, and specialized rooms. Instead of showing their back to the street and facing inward on courtyards, they turned around to face the street, and new Spanish buildings were set back from the street with an Anglo-style front yard. The gradual specialization of rooms was complete when plumbing arrived in the 1920s and established once and for all which room was the kitchen.[14]

---

[14] For a detailed analysis of the Anglo/Spanish hybridization see Christopher Wilson, "When a Room Is the Hall," *Mass* (Summer 1984), pp. 17-23. Good illustration of the Anglo/Spanish intersection is in Bainbridge Bunting, *Of Earth and Timbers Made* (Albuquerque: Univ. of New Mexico, 1974).

144

Timothy H. O'Sullivan. Museum of New Mexico. Neg. no. 72624.

1873 - By the time of this photograph the Zuni Pueblo complex was renowned for being five stories high, but it had been perhaps seven stories high in the early 19th century. The village was founded about 1400 AD. When the ladders were drawn up, the complex of several hundred rooms became a fortress against raids by Navajos and Apaches. New rooms were added casually to the top of the structure, turning former roofs into floors and terraces. The terraces were public walkways and the site of family activities such as food preparation. These photos were taken from atop the Brain Kiva, still in use, and the Big Plaza in the foreground continues to serve ceremonial occasions such as the *Shalako* celebration in mid-winter.

18 September 1882. Ben Wittick. Museum of New Mexico. Neg. no. 16054.

1882 - A party of visiting whites provides entertainment for the Zunis. Ground-level doorways are beginning to appear.

The core structures of Zuni pueblo dispersed as the need for defense diminished and wheeled vehicles made streets useful. These maps by Victor Mendeleff, Alfred Kroeber, and Perry Borchers are reprinted from the excellent paper, "Contemporary Zuni Architecture and Society," by T. J. Ferguson, Barbara J. Mills, and Calbert Seciwa in *Pueblo Style and Regional Architecture* (New York: Van Nostrand Reinhold, 1990), pp. 103-121. North is up. I have added lines to show the angle of view in the photo series.

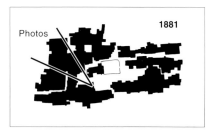

"CONTINUAL FLOW, continual change, continual transformation," said cultural historian Rina Swentzel, describing her own Pueblo village of Santa Clara in this century. The best documented transformation was at ZUNI PUEBLO, west of Albuquerque, New Mexico. For all the violent physical change, Zuni culture and traditions remain remarkably intact. The enormous changes evident in these photos were wrought piecemeal, family by family.

A. C. Vroman. Seaver Center for Western History Research, Natural History Museum of Los Angeles County. Neg. no. V-868. These two photos are in Vol. 9, *Southwest*, of the *Handbook of North American Indians* (Washington: Smithsonian, 1979), pp. 486 and 488.

ca. 1900 and **1978** - Zuni house interiors changed as radically as the exteriors. The 1900 living room shows whitewashed walls, a hooded fireplace, ledge for seating and shelving, a skylight, and kerosene lamp (right). A hooded fireplace still dominates the 1978 living room, where Francine Laate is about to make bread and G. Olaweon and Tom Awalate chat on the sofa.

Frederick Maude. Museum of New Mexico. Neg. no. 95320.

A. C. Vroman. National Anthropological Archives. Smithsonian Institution. Neg. no. 2293-B.

ca. 1895 - Sash windows brighten interiors, and the coming of wagons starts to open up the village with streets.

1899 - Sandstone masonry and a higher ceiling probably accompanied the foreground building being used for the masked giant *Shalako* dancers.

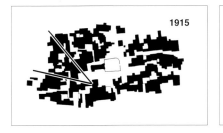

1915

1972

Museum of New Mexico. Neg. no. 61740.

ca. 1912 - More windows, more doors, higher ceilings in this photo by Jesse Nusbaum.

Museum of New Mexico. Neg. no. 5019.

1945 - Even newer masonry, soon to be followed by pitched roofs.

Melga Teiwes. Zuni Archaeology Program, Pueblo of Zuni.

1992

9 March 1992. Brand.

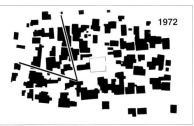

1934 - At Acoma Pueblo (west of Albuquerque), the house of Indian Santana Sanchez radiates the textured dry austerity that became the Santa Fe aesthetic, but his *fogon* is highly functional—cooking food, heating the small room, providing a convenient shelf, and staying out of the way of traffic in the corner. The practice of hanging working stuff on the walls—Shaker-fashion—was not adopted by Santa Fe style.

ca. 1935 - The house of artist Randall Davey in Santa Fe exhibited the growing inventory of Santa Fe style decor items, including an added-on kiva fireplace. The building originally was a sawmill, built in 1847 by the US Army. In 1935 Davey had lived there for fifteen years.

1985 - An illustration from the nationally influential book *Santa Fe Style* (1986) shows a recent prestigious second home outside of Santa Fe. The tiled floor, Taos drum, leather chair, and antlered skull (made iconic by local artist Georgia O'Keeffe) have joined the decor canon. By now the room is designed around the kiva fireplace.

Goods from the money economy came slowly to the self-sufficient Pueblo Indians but eventually overwhelmed them. Since the need for defense was reduced under colonial protection, ground-level doorways began to appear. Some families began to move away from the massive "apartment" blocks to live closer to their crops and flocks. Glass windows replaced the old tiny windows of selenite (crystallized gypsum), and stovepipe replaced adobe chimneys. Adobe walls gradually gave over to stone or, later, concrete block. Leaky flat roofs—no surprise—were covered with low pitched roofs. Steadily since the 1950s the Indians have dispersed into subdivision-style housing, much of it funded and designed (often obtusely) by the US government. Through it all, they succeeded in maintaining their spiritual practices and cultural identity.

WHOSE FIREPLACE is it? Anglos call it a "kiva fireplace" and put several in every Santa Fe style building, carefully burning piñon firewood upright in the approved local manner. The name shows that Indians are considered chic and Spanish not-so-chic, because the built-in corner fireplace is in fact Spanish and is called a *fogon*. The Indians picked it up from the Spanish colonists. They never did put any in the kivas (ceremonial clan rooms)—that would have been too much of a cultural trespass.

At the very time the Indians were buying into Anglo convenience for housing, the Anglos were heading the opposite direction. In 1912 Santa Fe realized it was facing a crisis. Bypassed by the railroad, it was steadily losing population. The only hope lay in attracting tourists, and tourists were flocking to see the pueblos and Spanish colonial buildings such as the old mission churches. Led by former easterners—archaeologists Edgar Hewett, Sylvanus Morley, and Jesse Nusbaum, and the artist Carlos Vierra—the Santa Fe Planning Board began the search for a "Santa Fe style" that would evoke visibly the town's considerable history. It had to be different from the newly successful Mission Revival style in California. Jesse Nusbaum and Carlos Vierra made extensive photographic surveys of pueblo and vernacular Spanish buildings and assembled them into an influential exhibition. An idiosyncratic warehouse in Colorado by architect Isaac Hamilton Rapp was identified as a perfect example of the new style. It was a romanticized copy of the Spanish mission church at Acoma Pueblo.

Quickly a number of public buildings on Santa Fe's plaza were built or rebuilt in the new Pueblo-Spanish-Anglo blend. They had pueblo set-back upper rooms and dramatic *vigas* (ceiling log ends protruding from exterior walls). They had Spanish *portales* with exaggerated corbeled columns, mission-style towers and balconies, and pseudo-Spanish tiled floors and decoration. Hidden inside were the Anglo construction and services. The buildings looked massive, apparently crafted of adobe soft-sculpted by hands and weather, but most were actually stucco on brick or wood frame.[15]

It was a brilliant concoction. Soundly researched, ably carried through, fiercely enforced, the new style turned Santa Fe into America's most coherent old city. The style responded perfectly to the romantic cherishing of vernacular simplicity that arrived in America with the Ruskin-Morris-inspired Arts and Crafts movement, and it fed on later infusions of romanticism that came with the artists' colonies of the 1920s and hippies in the 1960s (I was in that group). Dominant Santa Fe style architects such as John Gaw Meem were able to stave off Modernism by claiming that their style incorporated Modernist principles. "Some old forms," he argued, "are so honest, so completely logical and native to the environment that one finds—to one's delight and surprise—that modern problems can be solved, and are best solved by [the] use of forms based on tradition."[16]

What about the tourists? In a 1912 speech the archaeologist/booster Sylvanus Morley had proclaimed:

> None of us may live to see the day, but sometime in the future there will surely come a generation of Santa Feans who will not be eternally sleeping at the switch; but who will realize the possibilities of a Glorified Adobe City, and reap the golden harvest therefrom. Then, and not until then, will Santa Fe enter upon the epoch of increased and ever increasing prosperity, which is hers by right of every association, historic, geographic, and climatic.[17]

In 1992, newspapers reported, "Santa Fe, NM, bumped San Francisco from the top spot in Conde Nast Traveler's annual Readers Choice poll as the best travel destination in the world."[18] A home-decor style book titled *Santa Fe Style*[19] was a national bestseller. All through the 1980s, "Santa Fe" shops and restaurants permeated America's malls and airports and even invaded Europe. The galleries of tiny Santa Fe became America's third-largest market for art, following New York and Los Angeles. Commercializing vernacular had turned adobe into gold.

---

15 Real adobe currently costs about 40 percent more than other forms of structure because it is so labor-intensive. The saying is, "You have to be very rich or very poor to build with adobe in Santa Fe." Adobe has one advantage in being absurdly easy to demolish: just let the rain get at it.

16 John Gaw Meem, "Old Forms for New Buildings," *American Architect* (145: 2627; 1934), pp. 10-21; cited in Christopher Wilson, "New Mexico in the Tradition of Romantic Reaction," *Pueblo Style and Regional Architecture*, Nicholas Markovich, et al., eds. (New York: Van Nostrand Reinhold, 1990), p. 185.

17 Sylvanus G. Morley, "A Most Selfish Thing for Santa Fe," quoted in Nicholas C. Markovich, "Santa Fe Renaissance: City Planning and Stylistic Preservation, 1912," title above, p. 205.

18 *San Francisco Chronicle* (22 Sept. 1992), p. D4.

19 Christine Mather and Sharon Woods, *Santa Fe Style* (New York: Rizzoli, 1986). It sold 110,000 copies in four years and inspired half-a-dozen imitators.

1900 - The church of San Estaban at Acoma was begun by Fray Juan Ramirez after his arrival as the mission padre in 1629. The construction labor took ten years.

ca. 1915 - An attempt at restoration in 1902 produced blocky bell towers. Earlier known restorations occurred in 1710 (when the bells date from) and 1810. The loggia in the forground is part of the convent attached to the church.

ca. 1940 - A group of Anglo preservationists called the Society for the Restoration and Preservation of New Mexico Mission Churches raised money to restore the church and put Santa Fe architect John Gaw Meem in charge of the project. The work ran from 1924 to 1930, employing Acoma labor. For structural as well as aesthetic reasons Meem designed a slight batter (slope) in the bell tower walls, which were rebuilt of stone set in adobe mortar.

1908 - It was all a client's idea, actually. Colorado businessman C. M. Schenk, president of the Colorado Supply Company, asked his architect to use the church at Acoma as the model for a warehouse at the company's mining camp at Morley, Colorado. The architect was Isaac Hamilton Rapp, who had done strictly conventional brick and stone buildings up to that point. He made a stab at what the San Estaban bell towers might once have looked like and reversed the position of the convent loggia so the towers would draw the eye of passing train passengers toward an arroyo to the right.

The archaeologist Sylvanus Morley was in charge of an exhibit to be called "New-Old Santa Fe" at the Palace of the Governors. On September 20, 1912, he wrote to Rapp:

> Quite by accident, there has fallen into my hands a picture of the Colorado Supply Co.'s store at Morley, Colorado, designed by you. The thing is so absolutely in the spirit of "The Santa Fe Style" that I am taking this liberty of asking you to allow us to exhibit the original drawings, maps, elevations, etc. of this structure at our coming exhibition.... The extension of the native architecture to all kinds of buildings is, I believe, possible; and your success at adapting an old church to the highly specialized needs of a commercial house confirms me in my belief." [Carl D. Sheppard, *Creator of the Santa Fe Style* (Univ. of NM, 1988), p. 77]

Rapp sent a watercolor, and the exhibit was a huge success. The building itself, kernel of so many others, was later demolished.

1992 - In any weather (I was there in bracing snow flurries), a visit to the mesa citadel of Acoma is one of the great Southwestern experiences, always climaxed by the Acoma guides with a visit to the historic church.

1915 - The New Mexico Building at the Panama-California Exposition in San Diego, designed by Rapp, was a sensation. So was the Painted Desert Exhibit—five acres of cliff ruins, Navaho hogans, and pueblos—assembled by Edgar Hewett's New Mexico Museum, which was learning ever more about attracting and educating tourists. (San Diegans liked Rapp's stucco-on-wood-frame building so much that they kept it, and it is there still, much remodeled.)

ca. 1919 - Of course Rapp got the commission to build the Fine Arts Museum of New Mexico (1916) in Santa Fe. It is a beautifully designed building, inside and out—its details an encyclopedia of contemporary research on Spanish and pueblo buildings. Santa Fe style had arrived. A complete exemplar was in place. Rapp had refined his bell towers through three iterations. Ten years later John Gaw Meem would add one more back at Acoma.

1991 - Rapp's third copy of the Acoma church is safe from demolition or remodeling. Like the Palace of the Governors *portal* across the street on Santa Fe's plaza, the Fine Arts Museum is built of brick and stucco and is so revered only a terrorist would dare change it.

IMITATION RESHAPES ORIGINAL. The massive adobe Spanish mission church (1630s) at Acoma Pueblo is a true Spanish-Indian hybrid and one of the most impressive buildings in the Southwest (left). By the late 19th century, the bell towers were so eroded that no one knew how they originally looked. One architect (Rapp) guessed at their appearance in a series of Santa Fe style imitations of the building (above). Then another architect (Meem) restored the original church so that it matched Rapp's imitations.

San Esteban was a prodigious 17th century undertaking. The largest of all Spanish mission churches in the region, it served the most remote population—a village on top of a sheer-sided 400-foot mesa. There is no water, no trees, no dirt. The church is 150 feet long inside, 33 feet wide, 50 feet high, with walls 10 feet thick. Its estimated 20,000 tons of adobe and stone (not counting the necessary water) had to be carried up the mesa on a steep trail. The 40-foot roof beams, tradition insists, were borne by hand from mountains 20 miles away. Indians were never good slaves. It must have been either faith or the joy of doing something really impossible and spectacular.

But native adaptivity got left behind when vernacular *form* was translated into "vernacular" *style*. Real adobe was inherently fluid, the opposite of stuccoed wood frame or concrete block. Chris Wilson summarizes the systemic loss: "It was the change from an ad-hoc, open-ended, accretion form to formal designs conceived completely before the building is built; from multipurpose spaces and widely shared design and construction knowledge to specialized rooms and uses in buildings and the specialized knowledge of professional architects and builders." Specialized knowledge distances buildings from users. Specialized space hinders future flexibility. Santa Fe style seized all that was picturesque in the local vernacular traditions and threw away much that was wise.

Popularity over time, such as Santa Fe style has achieved, is something worth exploring. What makes some building forms proliferate more widely than others? The question has also been raised in biology, where it is termed "hyperdiversity." What makes creatures such as ants, rodents, and orchids so common and variable? It might be worth examining the attractions of three house types that became hyperdiverse in this century—Cape Cods, bungalows, and mobile homes. Does their extended popularity say anything about what current vernacular evolution selects *for*?

Children draw houses as unpreventably as they draw faces. No matter where they actually live, they nearly all draw the same house—one story, door in the middle, two windows to each side, pitched roof seen from the front, a central chimney with a swirl of smoke, and an inviting path up to the door. The classic Cape Cod house. It is so simple, rudimentary, austere, and yet practical that it fulfills the mythic image of house.

In its original profusion in coastal Massachusetts (including Cape Cod) the house was built by people dealing with very little money and a lot of wind. Its three major downstairs rooms and upstairs garret huddled around the massive heat-storing central chimney serving multiple fireplaces. The house hunkered low to the ground with shingled or clapboarded walls and narrow eaves.

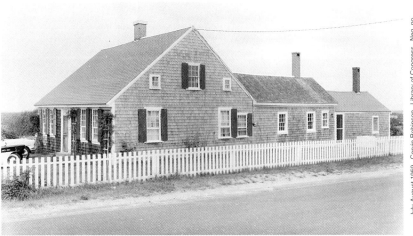

July-August 1959. Cervin Robinson. Library of Congress. Neg. no. 305675 - HABS MASS 1-THR 49-2.

1959 - CAPE COD HOUSE. In fact on Cape Cod (North Truro), this late-18th-century Cape has grown a sequenced pair of kitchen ells, flagged by the narrow stove chimneys. ("Ell" as in one limb of an "L".) Until the mid-20th century, kitchens were considered noisome necessities to be pushed as far away as conveniently possible. Apart from the formally symmetrical front, windows appear to have been added as needed.

It was so compact and solidly built of timber-frame that Cape Cods frequently were dragged on skids from site to site. "The most distinctive feature of the Cape Cod house is the roof," writes Stanley Schuler in *The Cape Cod House*.[20] It dominated the house, medium-pitched—8 to 12 inches to the horizontal foot (35° to 45°)—devoid of interruption or ornament. The roof's simplicity made it leak-free, cheap to build and maintain, and easy to add to. "Everyone started with the Cape Cod box and expected to add on," says Massachusetts builder John Abrams. "It's very easy to take those broad expanses of roof and put something into them— put in a dormer, have another roof come off for an ell. It's even easier to add onto the gable end, because you don't have to connect the roofs. Most modern buildings have their roofs much more broken up."

The Cape Cod was the standard cheap New England house from 1750 to 1850, spreading into New York and the Great Lakes states with a Greek Revival skin but the same essential floor plan. After decades of eclipse by Victorian and other styles, it suddenly re-emerged on a national scale. An editor of *Architectural Forum* wrote in 1949, "Twentieth Century America's most popular house design, now scattered throughout the entire country, is the Cape Cod Cottage."[21] Starting in the 1920s, a Massachusetts architect named Royal Barry Wills revived the low-cost Cape Cod house and found ready customers during the Depression. With larger windows all around and dormers upstairs for more room, the design flourished through the 1930s and 1940s and then took off in the housing boom after World War II. "That's when the Cape became the most widely built house in the world," says Schuler. "Its enduring quality is its simplicity, attractiveness and its basic integrity."[22]

What are the traits, then, that selection preferred? The Cape is small, solid, simple, cheap, growable, and carries a big roof. It looks respectable but stands apart from fashion, secure in the conservatism of its Yankee vernacular background.

If the Cape Cod house remained surprisingly true to type, a house form which became known for its diversity was the bungalow. It all began with a common epiphany. You're in your third week at some rude holiday cabin, luxuriating in its Spartan simplicity, its raw wood and stone, its casual access to the outdoors, and you realize you're shockingly happy. "This is so great," you sigh, "why don't we live like this all the time?"

Bungalows came to England originally as vacation cottages in the late 19th century, a northern-summer translation of what had become the standard rural house for British colonial officials in India (hence the exotic name). Hence also the name of its broad porch—verandah. The most distinctive feature of a bungalow, its low-pitched wide roof extending out over the verandah, was originally designed to deflect tropical rains and welcome tropical breezes. In Britain and later in America it served a different function, reconnecting the inhabitants to the outdoors. In both countries, quantities of bungalows were built for the growing middle class, which had acquired money and leisure enough for a rustic second home. Soon the same design was discovered to serve admirably on small city lots in the new streetcar suburbs with their new customer—small nuclear family, no servants, not much cash, wants a yard.

In America after the turn of the century, the bungalow came to represent a whole philosophy that was best expressed in Gustav Stickley's influential magazine, *The Craftsman* (1901-1916). Bungalow chronicler Anthony King writes that the magazine, "fusing architecture with social reform, was devoted to the development, both in theory and practice, of the three main principles of the Arts and Crafts philosophy—simplicity, harmony with nature, and the promotion of craftsmanship. The bungalow was to become the incarnation of all three."[23] The physical expression of this philosophy reached its pinnacle in the work of California architects Greene and Greene, whose sometimes extravagant bungalows in Pasadena (1904-1909) are venerated to this day.[24] The combination of an appealing philosophy, high-style fame with Greene and Greene, and the quick-and-cheap needs of new subdivisions made the "California" bungalow the standard expansion housing of the 1910s and 1920s. Sears & Roebuck sold them by mail order. Architectural historians regard them as the direct parent of the ranch house, which spread bungalow horizontality even broader on the bigger lots of the automobile suburbs of the 1940s and 1950s.

20 Stanley Schuler, *The Cape Cod House* (West Chester, PA: Schiffer, 1982), p. 13. Additional historical lore can be found in Ernest Allen Connally, "The Cape Cod House: an Introductory Study," *Journal of the Society of Architectural Historians* (May, 1960).

21 Quoted in Stanley Schuler, title above, p. 15-16.

22 Quoted in Clare Collins, "Old Houses Are Tremendously Strong," *New York Times* (9 April 1989).

23 Anthony D. King, *The Bungalow* (London: Routledge & Kegan, 1984), p. 134. King's subtitle is: "The Production of a Global Culture." He makes his case, since "it is a dwelling type—possibly the only one—which, both in form and name, can almost certainly be found in every continent of the world." (p. 2.)

24 The exquisite oiled woodwork, rough stone, and grandly connected rooms in houses by Charles and Henry Greene continue to inspire contemporary builders, clients, and some architects. The best books are *Greene and Greene*, vols. 1 and 2, by Randell L. Makinson (Salt Lake City: Peregrine Smith, 1977 and 1979). Vol. 1 is "Architecture as a Fine Art"; Vol. 2 is "Furniture and Related Designs."

1920 - At the South Florida Fair in Tampa, a model bungalow promises a quick ride on Florida's real estate boom—"Completed in 7 1/2 Days With 5 Men."

Bungalow scholar Clay Lancaster claims that the bungalow let America escape from the dead weight of Victorian and Queen Anne style:

> The bungalow vogue made new and definite contributions to the evolution of home planning in the direction of informality and unpretentiousness, use of common, natural materials, integration of house and landscape setting, simplification of design that became closely allied to practical requirements, and concentration on livability.... The American house during the bungalow period became lighter in construction, more flexible and open of plan, and less fussy in its furnishings.[25]

The bungalow, in other words, was an exceptional teacher.

People who live in those old bungalows today say there's not much occasion to remodel or demolish them because they work so well as is. They're a little cramped and dark, but the open

BUNGALOWS were mass-marketed by lumber companies ("From the Forest to You") as well as individually crafted to a high level of sophistication by architects such as Greene and Greene (see page 68). In the judgment of architectural historians, "The bungalow is one of the most successful vernacular houses ever built. It has been adapted for all regions and all climates. It was built in clusters, in rows, and as single houses, finished in several aesthetics, and scaled up and down both as to size and cost." [Jan Jennings, Herbert Gottfried, *American Vernacular Interior Architecture 1870-1940* (New York: Van Nostrand Reinhold, 1988), p. 342.]

layout of rooms with space-saving built-in benches and inglenooks keeps them comfortable. (Bungalows were the first houses in America with a spacious "living room," a clever kitchen, and a porch built in.) The big overhanging roof protects the whole structure from rain and sun. The emphasis on detailed craftsmanship continues to reward the eye and fend off the repairman. When they were built, bungalows were referred to as "the least house for the most money," but it was a good investment, evidently.

If, on the other hand, you want the most house for the least money, a mobile home is what you get. Mobile homes typically cost a fourth to a half of what comparable site-built houses cost. That's why 10 percent of all houses in America are mobile homes, housing 12.5 million people. In 1985, mobile homes comprised one-fifth of all new houses sold in the US, and two-thirds of all new low-cost single-family houses.[26] Polite culture only notices them after a tornado or hurricane tears up a few, but in the words of the authoritative *Field Guide to American Houses* they are the "dominant folk house of contemporary America."[27]

This tenth of all US housing has only one book about it, fortunately a good one. Allan Wallis's *Wheel Estate* begins, "The mobile home may well be the single most significant and unique housing innovation in twentieth-century America. No other

innovation addressing the spectrum of housing activities—from construction, tenure, and community structure to design—has been more widely adopted nor, simultaneously, more widely vilified."[28] They began as travel trailers in the 1920s, creatures of the American highway. They grew gradually in size till they became "house trailers" used for temporary postwar housing, sometimes as long as 55 feet, but still only eight feet wide. One innovator, Elmer Frey, invented the term "mobile home" and the form that would live up to it, the "ten-wide"—a ten-foot-wide real house that would usually travel only once, from the factory to the permanent site. For the first time there was room for a corridor inside and thus private rooms. By 1960 nearly all mobile homes sold were ten-wides, and twelve-wides were starting to appear.

A mobile home is an instant house. You wheel it in one day, hook up to the local utilities, and you're home. Everything works—plumbing, wiring, heating. It was all assembled in one smooth operation at a factory out of light wood frame on a steel chassis, clad with aluminum sheeting. The roof of white-enameled metal reflects the sun and sheds rain better than most site-built roofs. Half of all mobile homes are in specialized parks, among the last real communities in America, drawn together in part by physical closeness, in part by the need for political solidarity against enemies.

Mobile homes are always being attacked. By aesthetes for their appearance. By bigots for housing the "wrong" people. By the construction industry for "unfair" competition. By local government for paying insufficient taxes. (In fact, mobile-home park operators usually provide services such as sewage, water, garbage, and thoroughfares that government is spared paying for.) Many counties simply outlaw mobile homes. In 1970 the federal government recognized that most of the nation's low-cost housing was in mobile homes, and it set out to help the industry, but it messed up. HUD (Housing and Urban Development) came up with such a burdensome set of regulations that small manufacturers were driven out of business. Allan Wallis points out that "the smaller manufacturers, who needed a distinctive

---

25 Clay Lancaster, "The American Bungalow," *Common Places*, ed. Dell Upton and John Michael Vlach (Athens, GA: Univ. of Georgia, 1986), p. 103.

26 These numbers come from the thorough study by Allan D. Wallis, *Wheel Estate* (New York: Oxford, 1991), pp. 13 and 230. See Recommended Bibliography. In 1988 the average site-built home cost $100,000 for 2,000 square feet ($50 per square foot)—$138,000 if you add the average land price. An average "single-wide" mobile home of 970 square feet cost $18,500 ($19 per square foot)—and land was either rented in a mobile-home park or semi-free on a relative's rural property. Even a really roomy multi-section ("double-wide") mobile home of 1,430 square feet cost only $33,500 ($23.40 per square foot).

27 Virginia and Lee McAlester, *The Field Guide to American Houses* (New York: Knopf, 1987), p. 475. See Recommended Bibliography.

28 Allan D. Wallis, title above, p. *v.*

27 April 1993. Brand.

1993 - TRAILERS, because of their mobility, lead surprisingly long lives. You can get an old one cheap, haul it somewhere rural where regulations aren't too strict, and start a homestead. The sequence here, near Willits, California, was started with the left hand trailer in 1980. The one in back came in 1982, and then the common room joined them. The day I visited, the place felt spacious and pleasant, full of books and sunlight.

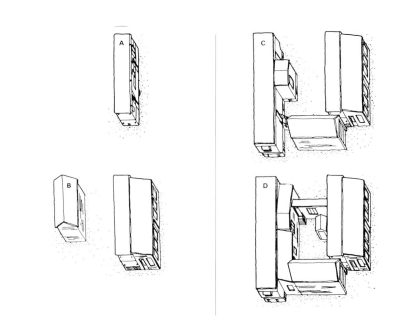

Assembling a mobile home compound has several advantages over just getting a bigger (double-wide) home. It costs less to start; it has better natural lighting; you can adapt better to family, financial, and site circumstances; and you get a nice enclosed courtyard after a while. This drawing of a typical sequence is by Allan D. Wallis.[29]

product to compete, were often a source of design innovation."[29] A highly creative industry was made stodgy by the strictures of approval.

But that hasn't yet stifled the creativity of mobile-home *dwellers*. The low initial cost and severe boxiness of mobile homes invites elaboration. Full-length shed roofs are added, first as a kind of porch, often later closed in to become new interior space. The need for storage, acute in mobile homes, customarily leads to the purchase of a "Tuff Shed" or other small metal structure. Sometimes, in unregulated rural areas, a whole normal house is built around the "seed" of a mobile home. More often, additional

mobile homes are added to form an informal compound, well daylit and loosely adaptive. Wallis writes, "The factory-built mobile home together with its site-built modifications represents an extension of two vernacular traditions, one industrial and the other user-based. In no small part the success of the mobile home as a form of industrialized housing must be attributed to the fact that it readily permits user modification."[30]

Wallis goes further. He quotes John Kouwenhoven on the attributes of American vernacular design—"resilient, adaptable, simple, and unceremonious"—and concludes that mobile homes express a native energy, an "aesthetic of process."[31] Mobile homes are openly make-do, unfinished. They embody the vitality and unembarrassed lack of dignity that J. B. Jackson sees enlivening the American future. They demonstrate vividly that however much

[29] Allan D. Wallis, "House Trailers: Innovation and Accommodation in Vernacular Housing," *Perspectives in Vernacular Architecture, III*, Thomas Carter and Bernard L. Herman, eds. (Columbia, MO: Univ. of Missouri, 1989), p. 42.

[30] Allan D. Wallis, title above, p. 41.

[31] *Wheel Estate*, p. 239. John A. Kouwenhoven, "What is 'American' in Architecture?" *The Beer Can by the Highway* (New York: Doubleday, 1961), p 156.

1991 - MOBILE OFFICES, just bare shells with a few windows and rudimentary services, are perhaps the most flexible and widely used of contemporary Low Road buildings. You see them used as film studios, classrooms, government offices, and on-site construction offices, as here on King Street in London. I confess that I looked at the highrise being built and wondered, "Why bother? Why not just keep stacking these things?"

16 August 1991. Brand.

buildings may be sold as a product, they are lived as a process.

So, what do the successes of Cape Cods, bungalows, and mobile homes tell about the vernacular process in industrial times? For one thing, vernacular is no longer regional, except in detail. (Mobile homes in really hot climates, for example, often grow an extra roof to ventilate away the radiant heat from the sun. In cold climates, pitched roofs are added to shed snow.) Successful building forms are broadcast nationally, driven by the national market economy. Builders and developers imitate the most successful of their competition. *That* is how buildings learn from each other in this century. Whatever buyers flock to will proliferate. What do they flock to?

It should be no surprise that Cape Cods, bungalows, and mobile

homes all are tiny. Small buildings are dramatically cheaper to build and to maintain. So long as people of modest means outnumber the rich, small will always win. Also, small invites the metamorphosis of growth. Only bungalows, because of their low overhanging roofs and tight city lots, discouraged growth and change. But bungalows were the most hyperdiverse of the three because of the quality of their legend. Popular songs were written about bungalows; whole magazines pushed their philosophy of informal naturalness. More than either Cape Cods or mobile homes, bungalows spread a pattern language—big homey living room with prominent fireplace, conveniences and furniture built in, connection to the outdoors, horizontality.

The difference between style and form is the difference between a statement and a language. An architectural statement is limited to a few stylistic words and depends on originality for its impact, whereas a vernacular form unleashes the power of a whole, tested grammar. Builders of would-be popular buildings do better when they learn from folklore than when they ape the elite. As for the elite: what might be accomplished with their abundant intelligence and creativity if architects really studied the process and history of vernacular designs and applied that lore in innovative work? We might get buildings that could be as original as needed, but still would feel profoundly familiar and right, and would invite change.

It would be a relief after all those smugly decorous buildings that "refer to" stylistic details of one vernacular tradition or another and miss the integrated lore. Of all buildings they are the most maddeningly perverse. They look like they should work, and don't.

CHAPTER 10

Function Melts Form: Satisficing Home and Office

MOST building *adaptation* is, like most building evolution, vernacular.

You don't have to look far to see it. The majority of people live in houses they own (64 percent of households in America, 66 percent in Britain).[1] The majority of workers work in offices (over 50 percent and still rising). Apart from high-turnover retail spaces, where do you find the highest rates of change within contemporary buildings? In owner-occupied houses and in office buildings. Whereas the remodeling of shops and restaurants is often the work of professional designers, the constant change in homes and offices is usually done by the occupants in a manner classically vernacular—informal, pragmatic, alive with offhand ingenuity, officially invisible. Direct, amateur change is the norm.

1986 - The Ragona family of Point Reyes, California. The idiosyncratic tower house with the great view of Tomales Bay had been owner-built in the 1970s in imitation of a lifeguard tower at a nearby beach. Tony Ragona recalls thinking in 1985, when he bought the place, "It was what we called in Philadelphia a 'trinity house'—three rooms above each other, with the kitchen on top. Only it was in the country." Right to left: Tony Ragona (40), his wife Virginia Drorbaugh (34), Travis (1), Seth (3); and Tony's sons by a previous marriage, Mark (10), and Steven (12), with the dog "Lady" between them. The cars are a '79 Volkswagen Rabbit and '72 Ford station wagon. (The photos are by Art Rogers—part of his renowned series, "Yesterday and Today.")

EVERY HOUSE IS A "BIOGRAPHY" HOUSE—like Washington's, Madison's, and Jefferson's—to some degree. Families can't help changing, and their homes can't help changing with them.

That's what was left out so badly in Modernist theory and still hasn't been corrected. A Modernist tract of 1940 stated:

> The essence of the new residential architecture is revealed in its twofold purpose: to base its plans upon the organic life of the family to be housed, and to make logical use of the products of invention. It has elected to make a fresh approach, to free itself of constraints, by consciously ignoring tradition and the expectations which the latter imposes with regard to facade and plan. *The outward form of the modern house becomes the outgrowth of a plan built around the interests, routine activities, and aspirations of the client and his family* expressed in terms of materials employed. Thus human need comes first. In skillful hands new appropriate and beautiful forms may emerge from an architecture which, discarding style, lets the house grow from the inside outwardly to express the life within.[2]

This "inside-out" design approach was thrilling, but it made the profound mistake of taking a snapshot of the high-rate-of-change "organic life" within a building and immobilizing it in a confining carapace—the expensive, low-rate-of-change Structure and Skin of the building. Too eager to please the moment, over-specificity crippled all future moments. It was the image of organic, not the reality. The credo "form follows function" was a beautiful lie. Form froze function.

It didn't matter. Life charges on and pushes mere material out of its way. Roofs can be raised; concrete walls are not *that* hard to

1992 - Six years later. The former kitchen on the top floor became the master bedroom. In 1987, Tony expanded with a high-ceilinged, shed-roofed addition on the left to make a combination living-room-dining-room-kitchen. He hired pros for the framing and did the finish work himself. The bottom floor became a studio for artist Virginia. The tower's original redwood siding wasn't holding up well, so in 1988 Tony put new plywood siding over it to match what was on the addition—adding vertical wood strips for a board-and-batten look. He installed double-glazed windows. Elsewhere on the four-acre property the family added a small workshop, a yurt-style second unit rented out as a bed-and-breakfast, a massage pool, and some thirty trees. The cars now are an '83 Volkswagen Jetta and an '82 Toyota pick-up. The people (and dog) are the same.

---

[1] In 1991 some 60.3 million of America's 94.3 million households were owner-occupied. The rest were rented, according to the US Census Bureau.

[2] James and Katherine Morrow Ford, *The Modern House in America* (New York: Architectural Book, 1940), reprinted as *Classic Modern Homes of the Thirties* (New York: Dover, 1989), p. 8. The italics are mine.

cut. The real action is all at the levels of Services, Space plan, and Stuff anyway. Function melts form. You can stupefy a building all you want and it will still learn.

Start with the home: such a melodrama. Within and behind the family situation comedy, the house itself is torn between fantasy and reality, driven toward change by the conflict. The house always aspires to be one thing and is obliged to be another. The fantasy was stated with suitable italics by Sir Henry Wotton in 1624:

> Every Mans proper *Mansion* house and Home, being the *Theater* of his *Hospitality*, the *Seate* of his *Selfe-fruition*, the *Comfortable part* of his own *Life*, the *Noblest* of his Sonnes *Inheritance*, a kinde of private *Princedome*; Nay, to the *Possessors* thereof, an *Epitomie* of the whole *World*: may well deserve by these *Attributes*, according to the degree of the *Master*, to be *decently* and *delightfully* adorned.[3]

In architect Sim Van der Ryn's modern translation, "Everyone who can afford it wants their home to be a nightclub big enough to entertain and impress everybody they know." High-end houses are vastly oversized and overpriced as a result, and low-end and middle houses are always installing tokens of immensity such as jacuzzi baths, coach lamps, foundation planting, and swimming pools.[4]

Fantasy palls, but reality sustains. One old house in Ireland inspired this comment:

> Home can be about architecture or a place in geography; or it can be about the sense of permanence we come to know through habit: an article of clothing repeatedly worn, a favorite turn of phrase, a melody of which we are fond, or the many visits to see a friend. Home is about the familiar, about gravity, about falling back into the self after being dispersed and overextended in the world.[5]

Far from being an Epitomie of the World and place of Selfe-

fruition, home is where you fall back into the self *from* the world, a place of honesty instead of aspiration, habit instead of ambitious striving. Returning to it you say with a sigh and double meaning: I'm home. (I've come home and I *am* home.)

The combat of extrovert fantasy and introvert reality was fought with fierce unconsciousness in many a 1950s home. Mom, driven mad by boredom and loneliness in her Epitomie of the World, shoved the furniture around, replaced the sofa and the drapes, and showed her accomplishments proudly to Dad, who came home exhausted from work to discover that his favorite chair had been given to the Goodwill. In time this process led to the bumper-sticker observation, "A man's garage is his castle." By the 1980s, married women had joined the workforce, and the favorite chairs of both sexes became safe.

You can find a vibrant combination of fantasy and reality in any

---

[3] From Wotton's *The Elements of Architecture*, quoted by Peter Thornton at the front of his spectacular work, *Authentic Decor* (New York: Viking, 1984).

[4] During the period 1960 to 1990, American homes grew by 50 percent to an average of 2,500 square feet at the very time that the average number of people in a house was diminishing from 3.4 to 2.7. Buyers apparently wanted houses not so much for themselves as for imagined future buyers, and architects didn't like the extra design work involved in smaller houses. "You need twice the design time and you get only half the fee," said one. Jerry Ackerman, "Changes Hit Families Where They Live," *San Francisco Examiner* (10 Jan. 1993), p. F1 (reprinted from *Boston Globe*).

[5] Andrew Bush, *Bonnettstown* (New York: Abrams, 1989), p. 12. Home reality at its best is portrayed in photo and word in this book, which records the fading glory of a well-lived grand old house near Kilkenny, Ireland. See Recommended Bibliography.

[6] Robert Dahlin, "Home is Where the Sales Are," *Publishers Weekly* (22 June 1992), p. 34.

[7] Carolyn Anthony, "A Book in Every Toolbox," *Publishers Weekly* (2 Nov. 1990), pp. 25-30.

[8] US Bureau of the Census, in *Housing Market Statistics* (May 1992) from the National Association of Home Builders, p. 30.

[9] David E. Nye, *Electrifying America* (Cambridge: MIT, 1990), p. 253. Electricity for the home was a feminist issue of the time.

[10] Estimates in 1993 claimed that 20 to 40 million Americans worked at home at least part-time—18 to 36 percent of the workforce, and that the number was growing at a rate of perhaps 14 percent a year.

untrammeled teenager's room, with its comforting clutter, passing hobbies, and garish posters—the room a sluice gate for the teenager's world. The mix is carried over into college rooms, architectural-student work areas, and increasingly, I would predict, in adult offices.

There's a fascinating fantasy/reality gradient in English country homes—proceeding from the always-public state rooms to the grand family rooms to the humble servants' rooms. It was the servants' quarters that turned out to be the most humane and easy to adapt, and many a member of the upper crust has retired gratefully to them, having turned over the front of the house to paying tourists.

Fantasy-based change in homes comes in blurts, but, except for the elderly, reality-based change is constant and relentless. Babies arrive, become kids, become older kids, leave; dependent aging relatives arrive, die; money comes, money goes; divorce hovers; careers change; everybody keeps on maturing in their tastes and activities. Meanwhile the world keeps tempting fantasy with niftier gadgets for entertainment, kitchen, and bathroom, and now that we've got a home gym and a sauna, how about a rock garden?

Do-it-yourself home improvement is a huge, growth industry. Back in 1980 Americans spent $44 billion on home improvement materials, tools, books, etc. And that was a fraction of the real economic event; the value of the unpaid labor involved—and the final improvement value—has never been estimated. Nevertheless by 1990, ten non-inflationary years later, the annual amount spent had nearly tripled to $110 billion, and the collapse of real estate would send it higher still as people replaced trading up with fixing up.[6] A book titled *The Router Handbook* (a router is a power tool for specialized shaping) sold 700,000 copies between 1983 and 1990—ten times greater than Witold Rybczynski's bestseller, *Home: A Short History of an Idea*. Go ten times greater still to 8 million, and you have the worldwide sales since 1973 for the *Reader's Digest Complete Do-It-Yourself Manual.*[7]

That's the amateur work. In 1980 professional home remodeling came to $46 billion (this number includes labor). By 1990 it was $107 billion.[8] There's an interesting reversal in those numbers—in 1980 more was spent on professional remodeling than on amateur; by 1990, amateur was ahead.

What was everybody doing with that money? A lot of it is just keeping up with the times. Ever since about 1880, changes in technology and society have put increasing strain on houses. Victorian houses were "industrialized" by the arrival of city water and gas, transforming kitchen and bathroom, and night use. When city electricity replaced gas (by 1920), small rooms that were kept closed to conserve heat and restrict the gas smell gave way to a more open plan. Architects weren't interested in electricity, but women and builders were—women for the convenience of use, builders for the convenience of construction.[9] Materials kept improving. Plate glass changed windows, concrete changed foundations, plywood and sheetrock changed walls. Coal furnaces converted to oil, and suddenly the basement was usable for workshop and rumpus room.

Families shrank, servants disappeared, the car arrived. Telephones changed the home's connectivity, and television changed it again—even demanding its own room. Economically, the home had been completely transformed in a hundred years from a place of production to a place of consumption, but that didn't hold still either. With the coming of personal computers and burdensome commutes, millions began using the home as an office, a place of production in the information economy.[10] And social change kept accelerating. Nuclear families exploded. Energy costs suddenly mattered. The whole society got older.

While trying to be a refuge from all this change, the home became its most fluid physical expression. People responded to the new needs and desires with direct, vernacular action. J. B. Jackson has noted, for instance, how the conversion of the home into a center for recreation and entertainment proceeded after the late 1930s:

John W. McCalley. Both photos from his *Nantucket Yesterday and Today* (New York: Dover, 1981), p. 38. See Recommended Bibliography.

PORCHES ARE PSEUDOPODS. Sometimes the household flows into them; sometimes they withdraw. Between growth pressure from within and weather stress from without, porches seldom last.

ca. 1915 - A late-19th-century summer house on the island of Nantucket (Massachusetts) appears to have already added a separate bedroom on the left and a wraparound porch on the right.

1975 - Sixty years later, there are signs that successive generations are sharing the summer house. Part of the wraparound porch has been glassed in, and half of the back porch has been absorbed into the house. Meanwhile, the gambrel-roofed back wing has acquired two more dormer windows upstairs, and the back bedroom has grown a second story. Three sets of wood steps rotted and were replaced. Somebody got tired of the shutters. Wood shingles disguised the surgical scars.

There is in fact scarcely a space in the modern American dwelling that owners themselves have not transformed in keeping with this new image. Even the backyard, freed of its clothesline and rubbish and of the obsolete garage, became a recreation area well before homebuilders saw its potential charm. Barbecue pit, plastic wading pool, power lawnmower, all antedate the developers' concept of Holiday Homesteads. And the garage as a family center half outdoors, part work area, part play area, is also a family invention, not the invention of designers.[11]

We begin to understand why site-built, platform-frame houses have persisted so long in America. Site-built is site-rebuildable, much more than factory-made housing, even mobile homes. Platform frame—2-by-4 wood stud walls raised a floor at a time—is an amateur medium. You can build or rebuild an entire house with a power saw and a hammer (I did so once in Nova Scotia). For reasons unknown—perhaps our frontier history—Americans revel in doing major home projects themselves, and so we stick with forms that give us that freedom.

Consider the life history of the porch, the most conspicuously

19 November 1921. Wesley Bradfield. Museum of New Mexico. Neg. no. 51927. The photo is in Sheila Morand's *Santa Fe Then and Now* (Santa Fe: Sunstone, 1984), p. 68.

1921 - Carlos Vierra was the first artist to move permanently to Santa Fe, New Mexico—hundreds came later. His photographic studies of Indian pueblos and Spanish colonial villages were a major part of the inspiration of Santa Fe style (Chapter 9) and of his own house at 1002 Old Pecos Trail. Partially funded by territorial senator Frank Springer to be a model of a Santa Fe style house, it was built by Vierra himself over several years (1918-1924). He was the first of countless artists in the region to fall in love with the scuptural qualities of adobe.

13 March 1991. Brand.

1991 - Vierra's three porches (called *portales* in New Mexico) didn't last long in the high mountain winters. The one at the far left rotted away, and the other two—middle and upper-right—were filled in. Still in use as a home, Vierra's house was his most influential work of art, prefiguring thousands like it in the region.

dynamic element of the American house. From the 1840s onward, porches became a highly popular add-on. They were the setting of many people's fondest memories, of summer evenings and lemonade, of a time when a whole town knew each other and said hello. The porch was an outdoor room, simultaneously intimate and public.[12] But its sheer exposure made it ephemeral. Rain and sun ate the wood floor and steps, the roof supports, and especially the wood piers. "Of all the woodwork exposed to the elements," says an experienced carpenter, "none is so vulnerable as the white pine porch railing. With the right combination of faulty detailing and wind-driven rain, a railing can be reduced to

[11] J. B. Jackson, "The Domestication of the Garage," *The Necessity for Ruins* (Amherst: Univ. of Mass., 1980), p. 109.

[12] Folklorist Henry Glassie tells how the porch also was a bond between rural and urban values: "In the rural southern United States the idea of the home characteristically includes an external social space, a wide porch or shady yard, where encounter is casual and informal and continual. When people from such settings move to northern cities they provide different environments with simple tests of their quality. If these southern people, most conspicuously black people, are packed into high rise apartment buildings, designed on analogy with buildings that please upper class urbanites so that they include private cubicles but not places for casual gathering, the people correctly rebel against the building, forcing it to decay rapidly, making it an unpleasant, alienated scene. But when the same people move into late Victorian houses in formerly bourgeois white neighborhoods, where houses are supplied with deep front porches and the shady streets before them provide places to gather, they lavish affection on their houses and recreate the easy feel of a rural community within a thoroughly urban setting." In "Vernacular Architecture and Society," *Vernacular Architecture: Ethnoscapes: Vol . 4*, Mete Turan, ed., p. 283.

shredded wheat in about eight years."[13] Porches tended to disappear or to be completely rebuilt and restyled every generation.[14]

Or they gradually became part of the house interior. First came screens, against the bugs. Then came glass, against the chill. Then came insulated walls, against the winter. And then the spindly piers gave out and had to be replaced with a real foundation. Each stage proposed the next, driven inexorably by climate. Open porches still thrive south of the Mason-Dixon line, but they are absorbed into the house everywhere north—fantasy surrendering to reality. The social climate also changed. The street became loud with cars and trucks, and passers-by diminished. Inside, air conditioning and television beckoned. The porch was functionally obsolete by the 1960s.

In its place came the deck, this time facing the back yard. A classic do-it-yourself project (a book titled *Decks* has sold 2 million copies since 1963), the deck turned out to be just as ephemeral as the porch. Rain and sun rotted even pressure-treated lumber. When atmospheric ozone depletion made people afraid of exposure to ultra-violet rays, the whole suntan rationale of the deck vanished. A revival of the covered porch seems assured.

A different sequence happened with the garage. How did it come to pass that most garages have everything *but* cars in them? Originally hidden behind the house in the backyard, garages migrated to dominate the front, swallowing half the facade of suburban ranch houses and their descendants. From occupying 15 percent of a home's enclosed space in 1930, they grew to enclose 45 percent by 1960.[15] Ostensibly this reflected the coming of multi-car families, but even more it was the covert correction of another Modernist mistake. Modern-house doctrine of the 1940s stated:

> Man's space needs within the home have…been reduced to a fraction of their former proportions, with the resultant elimination of attics, sheds, storage cellars, work rooms, sewing rooms and laundry. Public provision of libraries, schooling, music, and recreation causes still further reduction of space needs for many homes. Easy access to shops reduces the size of storage space—closets, pantries—and kitchens.[16]

Easy access to shops, far from reducing material in the house, crammed the house with new stuff. The trend started by bungalows of having no cellar or attic continued with ranch houses, but the need for storage kept ballooning, and eventually all the spare stuff wound up in the garage, pushing the car back out on the street—which it didn't mind, being more weatherproof than the house anyway. (Heaped boxes also filled many a glassed-in back porch.) Along with the off-season sports equipment, Christmas decorations, and memory-filled clothing and bric-a-brac, the garage inherited the activities that used to grow into basements and attics—backup refrigerator, washer/dryer, kids' room, hobbyist bench, male retreat, spare bedroom (eventually provided with a kitchenette and outside door and rented out). The only garages still used for cars are where street parking is impossible or forbidden.

The pattern is clear. When the family looks around for some place to expand, which it always does, the easiest, cheapest, and quietest direction (no building inspectors, please) is into existing "raw" space whose initial function is deemed dispensable—the porch and the garage. Maybe it's time to bring back the attic.

---

[13] Scott McBride, "Railing Against the Elements," *Fine Homebuilding* (Nov. 1991), p. 68.

[14] "The most common means of updating the appearance of a house is to add, remove, or alter a porch." Virginia and Lee McAlester, *A Field Guide to American Houses* (New York: Knopf, 1984), p. 14. See Recommended Bibliography.

[15] Title above, p. 31.

[16] James and Katherine Morrow Ford, *The Modern House in America* (retitled *Classic Modern Homes of the Thirties*), (New York: Dover, 1940, 1989), p. 10.

[17] K. C. Leong, *San Francisco Examiner* (21 April 1991), p. F11.

[18] Lorrie M. Anders, *San Francisco Examiner* (7 June 1992), p. F11.

Houses evidently need more low-definition space for later expansion, and it's easier to add in than add on.

In or out, remodeling a home is a royal pain. Whether you move away during the work or try to share the house with the workers, everything is in a maddening uproar. It's horrifying to see your secure world dismembered—rip, bang, and a whole wall of your life is gone, strewing disconcerting entrails of wire. Dust and mess invade the rest of the house. Life becomes endless negotiation—with the contractor, the workers, and among family members. Some of the new ideas don't satisfy, and each change forces other changes.

It is for this reason, says remodeling contractor Jamie Wolf, that 87 percent of remodeling work comes by personal referral. "It's about trust," he explains. "You're in their *home* for six weeks to six months." When the process goes bad, it can be life-warping. One family near San Francisco had a second-story addition drag on beyond two years, so they called around to other customers of the lethargic contractor. "We were all so exasperated that we formed an ad hoc support group, visiting one another's unfinished homes and sharing spine-tingling horror stories only remodeling veterans can appreciate."[17] A professional mediator in Berkeley reported that one of his clients "burst a blood vessel in her eyes just talking about her contractor."[18]

The ordeal of major remodeling makes many people opt for continuous amateur adjustment instead, which has its own problems. Many a remodeling contractor has to announce grimly higher costs upon discovering the product of previous do-it-yourselfers—dangerously inept wiring, nowhere-near-code plumbing, sloppily installed windows and doors. All that work has to be redone along with the new work in order to get by a building inspection.

Even professional remodeling takes a toll on a building when repeated often enough. San Francisco plumber Rufus Laggren comments wryly, "I've seen some beautiful buildings in Pacific Heights that turned out to have 'learned' so much from the last

three remodels that the bathroom was about to fall into the living room, because so little was left of the structure after all those grand changes had been cut in. Folks move bathroom fixtures around, which means moving pipes around, which means drilling or notching more holes in the floor joists. Each time a toilet moves, a 3-1/2-inch pipe needs to follow it. Since the folks on the hills have more money to express themselves with remodeling, one usually finds this sort of problem in the best neighborhoods." It's a good example of the destructive power of money and of the way differently paced parts of a building can tear at each other.

Most painful of all is when the expensive improvement turns out to be worse than what it replaced. In recent decades many older houses in the American south that used to be up on wood or masonry stilts were decreed by their owners to look "poor" that way, so the open sides were boarded in. This cut off ventilation—the original reason for the stilts, forgotten when the local vernacular pattern language died—and termites and rot quickly brought the buildings down. Another frequent error is installing built-in furniture. It works beautifully when the goal is to save space, as in bungalows, but it petrifies rooms around one ephemeral function. Soon enough you have to demolish the walls to move the furniture.

The trick is to remodel in such a way as to make later remodeling unnecessary or at least easy. Keep furniture mobile. Keep wiring, plumbing, and ducts accessible. Because at some point sooner than you imagine your present arrangement will become suddenly and profoundly intolerable, and you will not be able to rest until you have a Corian island in your kitchen, or a walk-in closet, or a fireplace in the bedroom, or another bathroom, or a skylight over the stairs, or an electronic home theater, or a private study isolated from the noise of the home theater, or whatever it is.

"Every house is a work in progress," writes David Owen:

> It begins in the imaginations of the people who build it and is gradually transformed, for better and for worse, by the people who occupy it down through the years, decades,

centuries. To tinker with a house is to commune with the people who have lived in it before and to leave messages for those who will live in it later. Every house is a living museum of habitation, and a monument to all the lives and aspirations that have flickered within it.[19]

Looking at the life of houses foreshortened in time like this, do any generalizations emerge? One might be: the improvements and corrections that people make have little to do with style and everything to do with amenities. Even fantasy items like a home workout room may reflect a fad, but they are not fashion statements. When homes are reclad with fake clapboard, fake stone, and the like, it's an apparent style change, but the main event is getting a lower-maintenance skin on the place. Low maintenance is an amenity. A garden, a bigger kitchen, another upstairs room: all these improve livability, not stylistic coherence.

Far the greatest rate of change comes right at the beginning, as it does with everything that lives. It starts before the building is even complete. People building a house—or any building—always complain that "finishing is *never* finished." There are several reasons for that. You're down to detail, and details are endless. Also you're down to where the building most interfaces with the people who will be living in it, and they discover that some important things were left out, and some ideas that seemed so sensible on the plans aren't going to work. Last-minute revision—the most important stage of tuning a house—comes just when time and money are shortest. Aggravated compromise is the order of the day.

Finally the work crew goes away and the occupants move in. Inhabitation is a highly dynamic process, little studied. There's a term floating around the fringes of biology that applies—"ecopoiesis": the process of a system making a home for itself.[20] The building and its occupants *jointly* are the new system. The dwelling and the dwellers must shape and reshape themselves to each other until there's some kind of tolerable fit. It takes time

and money that are seldom budgeted for. A building can be stillborn if it is too thoroughly finished and fitted out and isn't given a chance to respond to the life moving into it.

Interestingly, a whole new ecopoiesis process has to be gone through each time there are new occupants. David Owen describes it perfectly in his house-fixing book, *The Walls Around Us*:

> I now believe that when a new family moves into a house, the house suffers something like a nervous breakdown. A few days after the deal is closed, water begins to drip from the chandelier in the dining room, a heating pipe bursts, and the oven stops working. The house is accustomed to being handled in certain ways. Then, suddenly, strangers barge in. They take longer showers, flush the toilets more forcefully, turn on the trash compactor with the right hand instead of the left, and open windows at night. Familiar domestic rhythms are destroyed. While the house struggles to adjust, many expensive items—including, perhaps, the furnace—unexpectedly self-destruct. Then, gradually, new rhythms are established, the house resigns itself to the change of ownership, and a normal pace of deterioration is restored.[21]

A building is an organizational device, which means it is a communication device, which means that a certain volatility is always carving away at the physical building, making sure that whoever cooks the meals can chat with the family while cooking

---

[19] David Owen, *The Walls Around Us* (New York: Random House, 1991), p. 5. See Recommended Bibliography.

20 The word comes from Robert Haynes of Toronto, who applied it to planting life on Mars. "Eco" is "home"; "poiesis" is "making." Pronounced "EEK-o-poy-EE-sis."

[21] David Owen, Title above, p. 7.

22 The word combines "satisfy" and "suffice". It was coined in 1956 by the distinguished systems theorist Herbert Simon: "Evidently organisms adapt well enough to 'satisfice'; they do not, in general, 'optimize.'" In 1958 he wrote, "To optimize requires processes several orders of magnitude more complex than those required to satisfice." *Oxford English Dictionary.*

(kitchen walls melt) or that the boss can be alone on the telephone (cubicle walls suddenly rise to the ceiling). Teenagers isolate themselves, accounting departments drift to the periphery—both to get away from the hourly crises of action central. Often information itself is a cheaper fix than physical correction. In the brand-new Music Building at the University of Michigan on a pair of double doors underneath a glowing (legally required) "Exit" sign is a hand-lettered warning, "NOT an exit!!!"

The countermanding sign is an example of the way most problems are handled in buildings once they're occupied. The solutions are inelegant, incomplete, impermanent, inexpensive, just barely good enough to work. The technical term for it, which arose from decision theory a few decades back, is "satisficing."[22] It is precisely how evolution and adaptation operate in nature.

Even after generations of satisficing, the result is never optimal or final, though it can be, like the random-feeling street lights of Venice, sublime. The advantage of ad hoc, make-do solutions is that they are such a modest investment, they make it easy to improve further or to tweak back a bit. This dynamic, of course, is what makes so many "temporary" buildings become permanent. Being conspicuously and cheaply adjustable, they can wrap around any use demanded of them.

"Convenience defines what gets done," observes software entrepreneur John Pearce. I might add that it defines how things get done. Porches and garages weren't designed for growing into. Families satisficed their way in, fixing one minor problem after another with the most convenient good-enough solutions.

Another advantage of satisficing is that it's usually done by the

**SATISFICING doesn't try to solve problems.
It reduces them just enough.**

All three photos 8 October 1993. Brand.

A triple satisficing is visible around the fax machine in my office (a derelict fishing boat). 1) To make a horizontal space big enough for it next to the old steering wheel, I hacked some of the ledge away with a saber saw. 2) But then sunlight from the south-facing window cooked the fax and made its display go dark, so I shaded it with the manual and a piece of paper. 3) When I opened the cabin doors for ventilation on hot days, wind would blow the paper away, so I weighted it with a lead-shot-and-leather paperweight that was sitting around. Zero expense, zero waste of time, scant investment in an evanescent piece of technology, but a highly customized and convenient workspace.

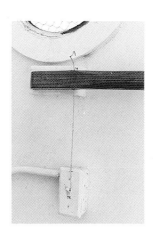

An incurable deck leak in the tugboat-houseboat where my wife and I live was making the kitchen unpleasant in the winter, because rainwater was collecting and dripping off a rusty carriage bolt in a deck beam and staining the back of the kitchen table bench. After two years of fighting it, we surrendered: a copper funnel under the bolt (right) collected the drip and conveyed it by copper tube to the outside. It looks rather elegant. End of problem.

When a carpenter friend replaced the rotted booby hatch on our tugboat, he inadvertently put the light switch too low, where one's hand grasping the rail (right) could no longer conveniently thumb the switch on one's way into the dark hold. "Fix it," we groaned. He did, very economically, with a piece of stiff wire.

occupants. Along with the efficient directness, you get a correct level of responsibility-taking. When local building adjustments are made by the local occupants, they "own" the improved space. The opposite happens when the occupants have to go through labyrinthine channels to get permission to make a change, and outside professionals are hired. Then the problem gets translated into one of interior design.

Interior designers never satisfice. They are paid to optimize, to make perfect. "All redecorations seem, when new, to be perfect and permanent solutions," warns interiors historian William Seale.[23] Perfection is frustratingly temporary always, but especially so in building interiors, where traffic is constantly buffeting the Space plan and Stuff.

In vernacular traditions, the inside and outside of a building were treated as one, but in the academic tradition they have been separated ever since the Renaissance, when sculptors specialized in the shape and exterior of grand buildings, and painters decorated the interiors. Over the last 100 years "upholsterers" and "drapers" became "interior decorators" and then, influenced by industrial design after the 1950s, "interior designers." They have been classic paraprofessionals, getting more work and often having more fun than many a licensed architect. The ones who focus on function—such as space planning for offices—have been largely a boon, but the ones who focus primarily on "look" have borrowed the artistic preciousness of academic architecture and made it even more restrictive by pressing it right around the people trying to use a building. Style cramps life, and life erodes style. All too soon the unified look is polluted by use, and it's time to hire someone to supply another alien unity.

Talking with William Seale, a professional consultant on historical interiors, I was surprised by the ferocity of his opinion of stylistic interior designers: "How many interior designers are trained? Very very few. They're merchandisers, and that's why architects claim they are not professionals. Most designers have something to sell, and they charge a percentage of it, and of course many take

kickbacks besides that. Many of the auction houses offer kickbacks running between 15 and 35 percent. They make a lot more money than architects. Their overhead is almost nothing." He added, "I think it's the most inconceivably boring business I could ever imagine, doing up people's houses."

It's also estranging for the clients to wind up dwelling in someone else's style package. Deborah Devonshire and her husband contemplated the issue when they were getting ready to move into the treasure house of Chatsworth:

> When I was young I watched my mother doing up whatever house we were living in and making it far prettier on far less money than those of friends who employed professionals to do the job, and I felt that I could probably do as she did, and so, for two reasons, we decided against employing a decorator. The first was that I cannot imagine living among someone else's taste, and the second that I cannot see the point of paying someone to do something I can do myself.[24]

Chatsworth feels more alive than many great houses because it embodies the intelligence and life history of the family living in it. One of the reasons visitors go back year after year is that even the most public and historic rooms are always somewhat different, reflecting the Duchess's evolving ideas.

And the private rooms? She writes of her sitting room with its out-of-fashion white paint and its Samuel Palmer watercolor (she thought the original was overpriced, and framed an illustration from the catalogue instead):

> Being in this room on a winter night, alone or with one or two great friends, the sparkling coal fire with its low brass-bound nursery fender, the familiar things all around, sitting in a chair which becomes a nest with letters and papers and baskets and telephone scattered on the floor, dogs comfortably settled by the fire, or near the draught of the door according to their thickness of coat, is my idea of an evening happily spent.[25]

The distribution of the dogs, and her perception of them, signal a room thoroughly grown into. Professional designers have borrowed all manner of Chatsworth fabric patterns and historical references and design inspiration, but they will never get the dog part right.

As Ivan Illich suggested, habitation and habit come from one word—Latin *habere*, to have. We shape our buildings around our routines, loving the fit when it becomes intimate and sure, and cleaving to it as conservatively as a duchess in her sitting room. Paradoxically, habit is both the product of learning and the escape from learning. We learn in order not to learn. Habit is efficient; learning is messy and wasteful. Learning that doesn't produce habit is a waste of time. Habit that does not resist learning is failing in its function of continuity and efficiency. Buildings keep being changed until they get to a point where they don't have to be changed so much.

Human learning has been distinguished into three fundamental levels, the first of which in fact serves habit. The keepers of a routine like to endlessly refine its detail, tweaking the environment to serve it ever more precisely. Far from threatening the routine, this kind of incremental refinement enshrines it always further. Organizational learning theorists such as Gregory Bateson and Chris Argyris refer to it as "single-loop learning"—like a thermostat turning the heater on and off to keep the room at a set comfort level. The learning is "single-loop" because it responds to a simple feedback loop: keep the room near 67 degrees Fahrenheit; keep the sitting room a restorative refuge for the duchess; keep the bathroom clean and cleanable.

The kind of learning that threatens habit is up a level—"double-loop learning." Instead of minor adjustments, major readjustments are called for. The thermostat is reset to a different temperature entirely. This is the second, higher, loop of feedback, declaring that the existing habit, no matter how perfectly refined, no longer serves the larger purpose. Sorry, Duchess, but the sitting room has to become a bedroom for the grandchild who is visiting more often now. Do you suppose you and the dogs could move to that nice corner room with the view toward the hills? Sorry, accounting department, you've outgrown the electrical capacity of this part of the building. We've got to recable, and how would you feel about a raised floor so the next change will be easier?

A third level of learning is "learning to learn." Raised floor is one example. This book is another. While single-loop refines habit, and double-loop changes habits, learning-to-learn changes how we change habits. An organization that fires the janitor and hires a facilities manager is taking a new relationship to change in the building, effectively shifting from a stasis manager to a change manager.

"Change is suffering" was the insight that founded Buddhism. We hate change. Ever since the big easy chair was reupholstered it's not as comfortable any more. And we love change. "Let's just *re-do* the kitchen!" To change is to lose identity; yet to change is to be alive. Buildings partially resolve the paradox by offering the hierarchy of pace—you can fiddle with the Stuff and Space plan all you want while the Structure and Site remain solid and reliable.

The best place to study buildings learning to take change as a constant—"learning to learn"—is where people do office work. Frank Duffy has remarked on the absence of long-view research in this area: "One of the most dominant facts of 20th century life is the huge increase in the importance of the office. We've had a massive change from less than 10 percent to over 50 percent of people occupied in offices, and no one has studied the physical evidence of that change from an organizational and social perspective." Particularly worth examining is the history of the

---

[23] William Seale, *The President's House* (New York: Abrams, 1986), vol. II, p. 1056. See Recommended Bibliography.

[24] Deborah Devonshire, *The House* (London: Macmillan, 1982), p. 79.

[25] Title above, p. 107.

"open office," an innovation that swept the world in a decade.

Few people realize that the open office—the scattering of desks and work groups around huge open floors—was a deliberate invention by a couple of professionals. Working near Hamburg, Germany, in 1958, the brothers Eberhard and Wolfgang Schnelle were neither architects nor planning consultants: they were organizational designers. And organizations, in their view, were severely restricted by tiny offices strung along lengthy corridors. Communication was poor, flexibility was nearly impossible, and the wrong-size groups were always stuck in the wrong-size spaces. So they threw away the walls and straight lines and created what was called *Bürolandschaft*—"office landscape"— later Americanized to "open office." It's hard to imagine now the shock and horror felt by orderly business people and architects when the idea first began to spread via magazine articles in the early 1960s. It looked like chaos, anarchy.

A second wave of innovation, in office furniture, completed the revolution. Robert Propst was known as an inventor of such devices as heart valves and timber harvesters when he was hired by the furniture firm of Herman Miller in 1960. By 1964 he was introducing ingeniously modular new work furniture, and in 1968 he wrote a revolutionary tract, *The Office: A Facility Based on Change*. Employing easily movable partitions and easily linked and delinked work surfaces and storage devices, the new furniture worked best in an open office environment. As advertised, it was relatively easy to move around and reconfigure for new jobs and new work groups. Form at last was following function.[26]

Furniture companies like Herman Miller and Steelcase prospered. "What's happened in the last twenty years is a massive migration of problem-solving from architecture into office furniture," observes Frank Duffy. "Things like dealing with storage, dealing with acoustics, with lighting, with partitioning, with cables have been taken away from architects." What used to be semipermanent Space plan material had turned into mobile Stuff. The driving force for all this malleability is what is called "churn rate"—the percentage of an office's population that changes location in a year. In most offices these days, churn ranges from 30 to as high as 70 percent—seven out of ten people physically moving within the organization every twelve months. As office work becomes ever less role-based and ever more project-based, the pace of shifting around work teams and players keeps increasing.

The final element making office environments forever liquid was the arrival of information technology. Office managers gradually realized that the computer equipment was going to keep needing

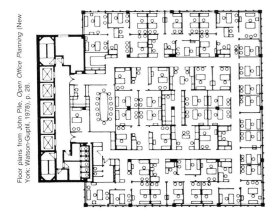

Floor plans from John Pile, *Open Office Planning* (New York: Watson-Guptil, 1978), p. 28.

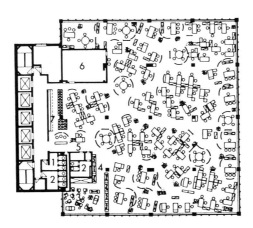

1967 - THE NEW OPEN OFFICE (right) was tested by DuPont in a comparison study with the company's standard cubicle office layout (left) in an office building in Wilmington, Delaware. The German team that had invented the form consulted on the project. Though it inspired many other companies to adopt open office space plans, the test did not persuade DuPont. Managers complained they did not have enough privacy, and no great improvement in costs or flexibility was detected. The movement took off anyway.

replacement, demanding major reshaping of work areas, every three years or so. Information equipment kept taking up more space and electricity, meanwhile displacing middle management, annihilating low-skill jobs, and increasing the use of specialists and technicians. "It's such a restless, remorseless, destructive technology," says Duffy, "it accelerates and destabilizes the whole process." It was the arrival of the permanent earthquake of information technology that pushed most organizations into hiring facilities managers, many of whom began as computer and communication-system techs.

At first, dropped ceilings took care of the growing traffic in communications wires. Recabling those wires could be handled by "poking through" the concrete floor plate into the ceiling space below, but that was soon found to be too expensive to change constantly, and so the raised floor once exclusive to mainframe computer rooms began to invade entire office buildings. At any churn rate above 30 percent, said the rule of thumb, raised floor became economical; it would pay for itself in two complete churns.

What is seldom noticed is that raised floor invites recabling *by the office users*, while dropped ceiling punishes it. The deeper the floor space, the better. In one intensely used computer-research room at MIT's Media Lab, I remember some students lifting floor tiles and diving bodily into the 20-inch floor space as often as three times a week to restring cable between work stations. Ceiling tiles, however, are an invitation to disaster. A typical account:

> There I was, standing on an old chair, sweating and cursing the moment I ever had the bright idea of stringing 100 feet of 4-conductor telephone wire from one end of my office to the other by running it through the dropped ceiling. I had

26 The history of the open office is traced in John Pile, *Open Office Planning* (New York: Watson-Guptil, 1978), pp. 18-36.

1969 - DROPPED CEILING and acoustical tile defaced the Great Hall of the Library of Congress's original Jefferson Building (1897) when temporary offices took over the space. One panel of ceiling tile appears to have broken and fallen out in customary fashion. This photo was used to lobby Congress to approve a third building for the Library.

forgotten about the lowly acoustical tile. This ubiquitous stuff, hanging over the heads of millions of American innocents who must be sheltered from the grim and boring reality of the building systems in their offices—heating ducts, BX cable, light fixtures, cable trays, and other effluvia—is crumbly, easy to break, difficult to cut and maneuver, and in general qualifies as a Building Material From Hell.

Pretty soon I felt it was going OK, although the stuff from the tiles was getting in my eyes and hair and all over the floor. I was taking it slow and easy—lift up a corner, push a little bit on the next corner, pop it up, sling the wire, run it along—

and then I ran into a veritable snakepit of cable bundles, light fixtures, and ducts, and one duct was flush with the center of the acoustical panel. I pushed too hard and, *fwuuuup*, the fucker cracked but good. A diagonal chunk about eight inches long popped off and disappeared into the drop ceiling. I couldn't get the rest of the panel to budge. I did the same thing to another nearby panel before giving up, sweaty and disgusted and dripping with acoustical panel drizzle. How did such awful stuff ever get to be standard in the construction, design and remodeling industries?[27]

It got to be standard because it was designed by professionals for professionals and assumed no incremental change and no handling by building dwellers. Liftable floor tiles also were designed by professionals, but incremental change by users was assumed. Ceiling tiles are anti-vernacular, floor tiles mildly pro-.

With the takeoff of the "information economy" in the 1970s and 1980s, new office buildings surpassed even malls in the real-estate boom. You would expect new commercial buildings to learn rapidly from all the changes going on with office use, and unfortunately they did. They overadapted to transient ideas, and colossal mistakes were made.

Highrise developers and designers were quick to welcome the idea of the open office. They soon discovered that open offices yielded a higher density of workers, hence higher space "efficiency" and higher rents. And all that openness permitted buildings to be much "deeper"—distances could be as great as 60 feet from the center of a floor out to the windowed edge of the building, again translating into higher efficiency.[28] Since floors were mostly all one space, lighting and heating/ventilating were greatly simplified. Just put in a fluorescent light every so many feet and a vent every so many feet and you were done. Hang a curtain wall of glass on the outside of your tall, fat box, and take a bow. (And hire some psychiatric specialists in the human-resources group to counsel employees suffering depression from loss of privacy and being cut off from daylight and weather.)

In 1973—the first energy crisis—all that exterior glass suddenly turned from a source of pride into an expensive problem. On sunny days too much sunlight was getting in, overloading the air conditioning, and too much heat was getting out the rest of the time, overloading the furnaces. So new office buildings were sealed tight, with tinted windows double- and triple-glazed, and tighter control was enforced by building management over lighting and air temperature. Money was saved, public credit was taken for energy conservation, and people became seriously ill from being sealed in with harmful chemicals outgassing from the carpeting and building materials, plus occasional pathogens breeding in the constantly recycled air conditioning. A new term entered the vocabulary: "sick building syndrome."

When information technology took off in the early 1980s, many of the highrise office buildings of the previous two decades were found to be absolutely incapable of adapting. Their floor-to-ceiling height was too low. There was no room for both a dropped ceiling and a raised floor, and no way to fix the problem. Blocks of new buildings in London's data-intense financial district were suddenly obsolete, and since the problem was Structural the only cure was demolition.

Some of the buildings that replaced them erred in the opposite direction. The "smart building" boomlet of the early 1980s was based on the idea of electronically integrating all the control systems of a building *and* offering tenants a full menu of built-in information services. Both failed. Climate control, fire suppression, security, lighting, and communications—all were

---

[27] Fred Heutte, in a June 1990 comment on The WELL.

[28] The gain was usually interpreted in terms of higher "net-to-gross ratio"—more rentable worker area per area necessarily dedicated to corridors, elevators, stairs, bathrooms, etc. Gross is the amount of square feet you have to build. Net is the fraction you can charge rent for. Efficiency of 90 percent is considered by landlords to be desirable, by occupants to be almost unbearable—galley-slave conditions.

1988 - THE NMB (now ING) BANK building (1987) in Amsterdam epitomizes the new trend in Northern European office buildings. Instead of one highrise core, it has fourteen low-rise cores joined by a spectacular interior pedestrian "street." Stairs predominate over elevators. Everyone in the building is within twenty feet of a window, all of which can be opened. There is little overhead lighting and no central air conditioning. The building includes Japanese roof gardens, four restaurants, and a snack bar for its 2,500 people.

supposed to be managed by a bank of computers tracking on time of day, day of week, who was in the building, and detailed sensing from all over the building. Integrating all the complexity in one bundle meant that only a specialist could understand or handle the system, and a problem in one area could infect the others. Seeking to improve control led to loss of control. One night at the headquarters of Bechtel, the world's largest construction firm, a group of senior executives met in the dark because none of them knew the phone code to turn on the lights.

Since all those pre-installed information services cost the developer about $2 a square foot extra, higher rents were charged. It turned out that no one wanted to pay extra, and "smart buildings" died in the market in just a few years. There was a contradiction at the heart of the idea, according to Steve McLellan, a telecommunications regulator in Washington state: "We found that any user sophisticated enough to seek out a 'smart' building was also sophisticated enough to home-brew a more flexible system." Tenants universally preferred to install their own communication systems.

What do the linked debacles of deep office buildings, sealed office buildings, and "smart" office buildings have in common? Each was a clever and comprehensive design solution, but each tried to solve just one primary problem and acted as if the problem would

hold still over time. These were classic cases of overspecificity, overcentralized control, and "tight fit." Each took a conspicuous trend of the moment—open offices in the 1960s, energy efficiency in the 1970s, information technology in the 1980s—and, at astronomical cost, shaped whole buildings tightly around it. When the trends moved on, the buildings were left standing, good at something that no one wanted any more. Their failure is the failure of optimization as a design strategy.

Revolution by the oppressed workers was predictable. Frank Duffy has the story of what happened to new office buildings during the 1980s in Scandinavia, Holland, and Germany, where open offices began. "Those big decisions made by architects and managers were superseded by *real* industrial democracy through the institution of workers' councils, which are critical in northern Europe. An employer has to consult the employees before he proposes a change in their qualitative work life such as a new building. And guess what they don't like? They don't like open plan. And guess what they do like? They like windows that they can open, and doors they can shut, and walls they can bang on. The new northern European buildings, instead of being 120 feet deep, are 30 feet deep. They're like hotels—millions of individual rooms, each with a window." The buildings typically are only five stories high, with lots of public circulation (70 percent "efficient").

The best of them, such as Ton Alberts's NMB (now ING) building (1987) in Amsterdam, are well-loved.

This change-back phenomenon is one I've observed so often in buildings that I suspect it approaches a law. Several dynamics seem to be at work. Change is often followed by reversal of the change, because the prior pattern lingers as the most conspicuous alternative, because people are understandably conservative about their physical space, and because most change is really undertaken as a trial, no matter what people say at the time. And most trials are errors. My photographic study of office furniture over time (in the Appendix, pp. 216-217) suggests that the most likely place for a piece of furniture to move after it has moved once is back to where it was before.

The best adaptation I've seen of the open-office idea is a partial retreat from it. People crave acoustic privacy so they can talk on the telephone, but visual privacy is not as important—they like being able to see what's going on. This has led to a very satisfactory compromise called "cave and commons." Each office worker has a private office, often small, which opens into a generous open area surrounded by many other private offices. The open area has a kitchen, some couches, sometimes tables for sitting around informally, and sometimes a working library, or at least a rack of current periodicals. You can shut the door of your cave and concentrate, or you can leave your door open and keep an eye and ear on who's coming and going in the commons, and whether the meeting or presentation going on there might be worth leaning in on. The feeling is congenial and homey, and it encourages the casual encounters which, research keeps showing, are at the heart of creativity in offices.

An extreme case of cave-and-commons is an extreme success. This is the Mathematical Sciences Research Institute (1985) at the University of California, Berkeley, designed by William Glass. A modest three-story wood building, its whole inside is one continuous atrium with just a rind of fifty-six small offices. These ascetic, contemplative caves for the visiting mathematicians all

8 July 1993. Brand.

1993 - THE ATRIUM of Berkeley's Mathematical Sciences Research Institute (1985) makes the building one connected whole. Every one of the 56 offices opens directly into the commons area, and crosswalks and stairs link them conveniently. The clink and chatter of daily afternoon tea drifts enticingly up from the floor of the atrium. But the building has become burdened by success. The programs have grown so much that all the offices designed for occupation by individual visiting mathematicians have been forced to double up, and the roommates disturb each other's concentration with phone calls, visitors, and chat— to the point that many now work at home, destroying the whole interactive purpose and glory of the building. A new wing is planned to add 30 new offices, a large commons room, a larger library, a cafeteria, an auditorium, and half a dozen small seminar rooms (sorely needed).

open directly onto the exposed walkways and stairways of the atrium. At the ends of every floor are comfy hang-out areas with couches and blackboards and a pleasant view. The mathematicians can't help encountering each other constantly, and there's the further mix of high tea every afternoon at 3:15 down on the

1993 - Deliberately designed to feel informal and non-institutional, MSRI's wood frame building with cedar siding is inexpensive and unpretentious. Architect William Glass avoided the temptation to do something geometrically or topologically cute and delivered a straightforward, serviceable building. His firm was rehired to design the new wing, which will expand to the left.

1993 - Well-used blackboards as well as pleasant views in the mini-lounges at each end of every floor encourage the mathematicians to convert chance encounters into searching discussions. In "cave-and-commons" offices, both privacy and easy access to other people are treated as crucial amenities.

floor of the atrium next to the library. Alumni of the building report that "Mathematics comes alive when people talk to each other." They hate to leave at the end of their terms, because their home universities have nothing like the building's collegiality.

Office buildings are organizational hardware, and since many organizations are redefining themselves these days as "learning organizations," the design question is: how can buildings aid organizational learning? One answer may well be: by aiding local adaptivity. Small groups adapt more quickly and accurately than large groups, and individuals are even quicker than that. Smart organizations, therefore, push control of space as far "down" the organization as they can. It starts with the kind of space that is leased. Rather than moving into offices already fitted out by some spec developer, adaptive organizations prefer either raw space in an old building or raw space in a new building. There it's called "shell and core"—only the exterior walls and the core of elevators, bathrooms, service chases, etc. are finished. The tenant takes active control of the space by filling in all the rest.

Once in place, the organization advances best by hordes of "small wins" in space adaptation rather than huge sweeping solutions that are two years out of date by the time they're finally agreed on and implemented. Chris Alexander and his students, in an unpublished 1990 work on "Office Patterns," define the physical spaces of an organization as a nested hierarchy of realms— individual, within workgroup, within department, within whole workforce, within the larger community:

> At each level of scale, it is those actually using the space who understand best how it can made/altered to have the character of being conducive to the work, and this group should be given sole control over that space both in the physical definition of the territory (including Realm Center and Realm Boundary) and by giving the group power over placement of furniture, purchase of needed items, decorations, etc. Thus an individual has control over his/her own workspace; the workgroup has control over the group working area but not over the individual workspaces; the department has control over its space but not over the workgroup spaces, and so on.

> Therefore we suggest using materials and structural systems which invite change and allow changes to accumulate,

2 December 1991. Brand.

1991 - The building at 5900 Hollis Street in Emeryville, California, was originally (1930s) a factory for International Harvester tractors. The space we moved into had more recently been a dance studio, then a school. I used the existing wood columns to hang the walkways for the second level of offices. Architect Philip Banta translated my rough space plan into a workable design, and a group called Dream Builders made it real.

24 February 1993. Brand.

1993 - The weekly staff meeting at Global Business Network takes place in the commons. Anyone having to miss the meeting because they're on deadline or taking an important phone call can nevertheless hear from their office what's going on. Each office has a solid door facing the commons and a window that can be slid open or closed. The architect had specced the walkway railings as solid sheetrock, but we realized they needed to be transparent to increase visibility of the offices and of people walking around. The receptionist (who sits by the entry just behind the camera) can see who's where most of the time. The space has a certain buzz like a newspaper city room, but people can be quietly alone with their work when they need to be.

MY DESIGN for a cave-and-commons office for Global Business Network was inspired by the Mathematical Sciences Research Institute atrium (previous page) and the "playroom" in the middle of the Artificial Intelligence Lab at MIT. In an old factory we built a two-story skylit commons surrounded by fourteen small offices linked by rather grand stairs in front of a wall of factory windows.

gradually fine-tuning some areas very closely to the real human needs that exist there. Other arrangements, for which the need might become obsolete, would disappear over time. (But the space that housed them might retain faint traces, a pentimento, of their previous use.)

In organizations and in buildings, evolution is always and necessarily surprising. You cannot predict or control adaptivity.

All you can do is make room for it—room at the bottom. Let the mistakes happen small and disposable. Adaptivity is a fine-grained process. If you let it flourish, you get a wild ride, but you also get sustainability for the long term. You'll never be overspecified at the wrong scale.

Have any office buildings proven adaptable over the decades on purpose? Some have managed by accident, such as the Chrysler Building (1930) and Empire State Building (1931) in New York. Their high ceilings, daylit shallow depth, and openable windows turned from embarrassments back into virtues without benefit of intent. The severely ecological architect William McDonough imitates them with his insistence that any new office building he designs be potentially convertible into housing, since he regards that as the most fundamental use of buildings, for which there will always be a need and which always guides you toward humane

A HOUSE WITH OFFICE-BUILDING FLEXIBILITY is the prototype "Next Home" from McGill School of Architecture in Montreal. Each 3-1/2-story unit can accommodate (and change for) one, two, or three separate tenants, and can be built as detached, semi-detached, or row housing. Like an office building, each unit has readily accessible (hence adaptable) Services chases running up the stairwells and horizontally on each floor. The building is designed to adjust easily to a variety of occupants—single as well as two-parent families, couples without kids, single tenants, and offices—and to be adapted by them.

design.[29] His ingredients? Modest depth, high ceilings, operable windows, massive construction, raised floor rather than dropped ceiling for services, and individually controllable amenities such as window awnings. "It won't do, any more, to think in terms of cradle-to-grave," he insists. "From now on we should think of our buildings in terms of cradle-to-reincarnation."

For me the most wholesomely, loosely, intentionally adaptive of white-collar work buildings is the Main Building at MIT. A five-story boxy sprawl on the banks of the Charles in Cambridge, Massachusetts, its claim to architectural fame is the "Infinite Corridor"—a 600-foot-long hallway of such width, traffic, and straightness that the one evening each year the setting sun shines

AFFORDABLE AND ADAPTABLE were the guiding principles for this no-frills and empowering approach to housing. It's an intentional Low Road building. Construction is simple, fast, and cheap—largely factory-built components assembled on site. At $US26/square foot to build, it can make owning a home cost no more than renting. The size is modest—20 feet wide by 40 feet deep; 800 square feet per floor. It's built to minimize operating costs such as energy and maintenance and to make remodeling inexpensive. Unlike mobile homes, the "Next Home" is well suited to an urban setting.

The rear view (right) of the prototype shows how the decks next to the fire stairs can convert to a porch or an enclosed part of the house.

Designed by Avi Friedman and colleagues at McGill's Affordable Homes Program in 1996, the "Next Home" was a follow-on to their successful 1990 project, the "Grow Home." It was only 14 feet wide, with a total of 1,000 square feet on two floors. By 1995 5,000 Grow Homes had been built in Canada as row houses, at a cost of only $US60,000 per unit.

---

29 The market supports McDonough's strategy. In 1993 an Urban Land Institute study showed that the most profitable use for the surplus office buildings left over from the 1980s real estate boom and crash was to convert them to apartments. Patricia L. Faux, *The Edge City News* (Vol. 1, No. 2), p. 8.

down its length is an occasion for university-wide celebration. Two things dominated its design in 1916. One was MIT's fifty years of exasperation with having been widely separated in buildings scattered around Boston. The other was an architect-hater named John R. Freeman, an hydraulic engineer by trade, who admired the design of New England's mills and factories and aimed to base MIT's new campus on their honesty, pragmatism, and massive connectedness.

Freeman spelled out a series of design principles that guide MIT construction to this day: "An abundance of window light and a flood of controlled ventilation with tempered and filtered air; Maximum economy in energy and time of students and instructors; Maximum economy in cost of efficient service in heating, ventilating, janitor service, and general maintenance; Maximum resistance to fire, decay, and wear; Maximum economy

in cost of building per square foot of useful floor space." So MIT's Main Building, which is still the core of the campus, is a web of high, narrow wings 64 feet wide—just right for a wide corridor in the middle, with space for a variety of classrooms, laboratories, and offices on each side. (A later MIT building of 55-foot width was found to be inflexibly restrictive.)

Flexibility is all-important because departmental space is reassigned constantly by the university. Some 5 percent of the university buildings change usage every year, which means a complete turnover every twenty years. Many areas started as laboratories, then converted to offices when technology obsolesced the lab, then became classrooms, then reverted to a new generation of laboratory. Overall the usage is 40 percent labs, 40 percent offices, 7 percent classrooms. This kind of flexibility is administratively possible because, unlike at other

MIT Museum. Neg. no. AE-83-01.

1983 - MIT'S MAIN BUILDING (foreground) is echoed by Building 20 (upper right). The 600-foot "Infinite Corridor" runs from the dome at the left, in front of the dome in the middle, to the right end of the building, where it forks left and right to connect to other buildings by enclosed walkways. The building combines exceptional flexibility, durability, and comfort with an urban intensity.

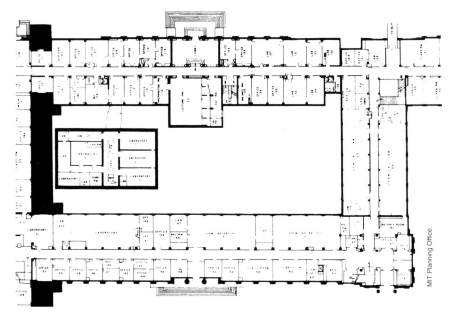

MIT Planning Office.

The offset corridors of the Main Building give further flexibility by offering different sizes of rooms on each side. New disciplines and technologies coursing through the building keep changing the usage, shape, and services of the rooms. This plan shows part of the east wing (right) of the Main Building.

campuses, departments do not have their own buildings.

Campus planner Robert Simha says that the primary benefits are intellectual stimulation and departmental evolution. "The thing which has characterized MIT's success is a physical environment which does not impair communication and set up arbitrary barriers to it. There are no boundaries, no locked doors, no signs that say this is mine and that is yours. You can wander unfettered from one discipline to another without even knowing you have, and bump into a physicist, then a few feet away bump into a chemist, then a mathematician. The action at MIT occurs in the public ways and intersections, just like in any town. We've seen that areas which are physical cul-de-sacs soon become intellectual cul-de-sacs."

Flexibly vague boundaries are of the essence, says Simha, because "we deal in science and technology. It's not a conservative activity. It's an innovative activity—things can expand and contract in an organic institution that is populated by people whose activities get born, live, die, and get replaced. Some things wither, and some stimulate other things." In Simha's view, funky old Building 20 (Chapter 3) is so successful because it is a direct copy in wood of the Main Building, whereas the Media Lab building "violates every principle that has guided MIT planning since 1916." It is isolated from the campus-wide warren of corridors, drastically inefficient in usable space, and inflexible in layout.

Current planning documents at MIT emphasize future building adaptivity based on: "loose fit"—generous dimensions that permit easy change of Services and usages without having to change Structure; robust construction, so that a present office can later become a lab with heavy equipment or even a library; and horizontality. This last comes from MIT's Thomas Allen, a researcher on the influence of space organization on innovation: "Both social research and experience at MIT demonstrate the benefits of buildings which are horizontally organized and low enough for stairs to be the predominant method of vertical circulation. It is estimated that communication between members of the same organization decreases by an order of magnitude [i.e., to one-tenth] when their offices are on different floors."[30]

Modern society's two great vernacular spaces, the office and the home, seem to be interpenetrating. Trends in office work (such as electronic "telecommuting" and corporate "outsourcing" of former in-house services, plus the continuing boom in tiny start-up businesses) are pushing more and more people to have a home office. Meanwhile, individual workers in office buildings are getting ever more control over their own work areas, which makes the offices increasingly homey. Both kinds of buildings thus become more fluid, as the real drivers of "learning" in buildings—the individual users—treat their buildings as an extension of themselves.

The common attribute of vernacular remodeling (and construction) is that it is done without plans. You proceed by improving on what already exists, following wherever usage demands. "Wanderer," wrote a Spanish poet, "there is no path. You lay down a path in walking."[31]

[30] *Main Campus Northeast Sector Master Plan* (Cambridge: MIT, 1989), p. 48. For a highly readable account of MIT's innovative environment see Fred Hapgood, *Up the Infinite Corridor* (New York: Addison-Wesley, 1993).

[31] *"Caminante, no hay camino, se hace camino al andar."* Antonio Machado, *Proverbios y Cantares* (1930). These lines translated by Francisco Varela.

CHAPTER 11

The Scenario-buffered Building

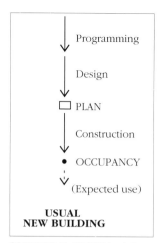
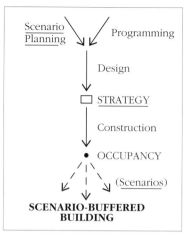

SCENARIO PLANNING leads to a more versatile building. It takes advantage of the information developed by programming (detailed querying of building users) and offsets the major limitation of programming (overspecificity to immediate desires). The building is treated as a strategy rather than just a plan.

ALL BUILDINGS are predictions.

All predictions are wrong.

There's no escape from this grim syllogism, but it can be softened. Buildings can be designed and used so it doesn't matter when they're wrong. Vernacular building types often have this quality of forgiveness, but how do you develop it in a new building, or in a difficult building that you're stuck with?

I suggest that there is a tool ready to hand, not used by the design professions before, that could be as useful to a home owner-builder as to a city planner. The tool, called scenario planning, has been evolving quietly for thirty years—first in a military context, later by corporations forced to think ten years ahead by a business environment which had become so turbulent that traditional forecasting was useless. The product of skilled scenario work is not a plan but a strategy. Where a plan is based on prediction, a strategy is designed to encompass unforeseeably changing conditions. A good strategy ensures that, no matter what happens, you always have maneuvering room.

Many architects would insist, "We already do that. It's the whole point of programming. We talk to the people who will be the building's users in great detail to find out exactly what the range of their future needs might be, and then we build around that." Programming is indeed one of architecture's great achievements, a sophisticated planning technique that might be profitably employed by many another industry. But it has limitations built in that are worth examining, because they make explicit a systemic fault in modern architecture. The cure also, then, must be systemic.

It was Viollet-le-Duc, the mid-19th-century French restorer and architect, who first encouraged designing a building around its analyzed needs. That made him the forefather of Modernism. By the 1920s, architects were using "bubble diagrams" to show anticipated traffic and proximity requirements—defining which uses should be near or remote from each other. The space plan and even the building exterior were then shaped around those diagrams. In the 1950s, the emergence of operations research in corporations inspired new levels of specificity in building "predesign," and the preparing of building programs became a formal discipline. (In Britain the program is called a "brief.")

The best description of programming technique is a perennially updated book with a pertinent title, *Problem Seeking*.[1] Noting that many a building is a brilliant (or pedestrian) design solution to *the wrong design problem*, the authors spell out a detailed procedure for working with the client and expected users of a building to

---

1 William Peña (with Steven Parshall and Kevin Kelly), *Problem Seeking* (Washington: American Institute of Architects Press, 1969, 1977, 1987). See Recommended Bibliography.

find out exactly what they want and need and can afford. You then state the design problem(s) in those terms. A product of America's first architectural corporation, CRSS, the book emphasizes intense teamwork at every stage, in defiance of the romantic tradition of the lone and lordly architect.

All too quickly the practice became just another shard in the fragmentation of the building professions. Specialized programming consultants nowadays are hired by an architect or a client (or each separately!) to begin the predesign process. Unfortunately, they are so expensive they're usually employed only once, so the problem-seeking process is given no opportunity to adapt.

The result of good programming can be a building that works uncommonly well. One reason I. M. Pei's Media Lab building (1986) at MIT works so badly is that the architect had to proceed without a program, due to prolonged departmental bickering about who would be in the building, and the fundraisers wanted to have a design early so they could sell it to potential funders. "I. M. was pawing the ground waiting to get the program, and no program was really forthcoming," MIT planner Robert Simha remembers. "When architects design buildings for themselves, you invariably have an interesting time."

Pei has said that academia often is a difficult client, and he's right.

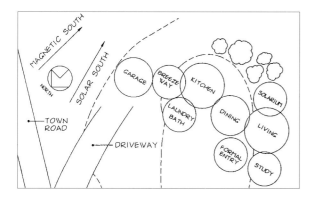

A BUBBLE DIAGRAM is a convenient graphic shorthand for showing how anticipated uses of a building might relate to each other in space. At the height of Modernist design, the bubble diagram was simply squared up and roofed. This sample diagram is from Tedd Benson, *The Timber-Frame Home* (Newtown, CT: Taunton, 1988. See Recommended Bibliography.)

At MIT the client was at first too vague and later too specific about its needs. Boston architect William Rawn notes that professors are notorious for wanting space designed precisely around their current research, and never mind the future. The Media Lab is burdened with a number of expensive overspecified spaces, such as a pair of rooms designed for wall-size rear-projection—research that was no longer even going on when the building opened. The rooms defaulted to poor storage space, then even poorer office space. A long, narrow theater was built and wired especially for advanced interactive movie research. All it's used for is lectures, and its design makes it one of the worst lecture spaces on campus.

A comparable research building that works very well, thanks to intelligent programming, is the Lewis Thomas Molecular Biology Laboratory (1986) at Princeton University in New Jersey. It was designed by Post-Modernist star Robert Venturi in graceful collaboration with a technical building specialist firm, Payette Associates. A published report on the project describes how an insightful program led to design decisions that provided precisely the amenities lacking at the Media Lab:

> Architect Jim Collins [job captain for Payette] learned that microbiologists are social people who feel strong ties to their scientific community…. The team's work was guided by this understanding of the users' communal spirit, as well as technical constraints. [Department chairman Dr. Arnold] Levine said that his highest priority in planning the building was "to force interaction between research groups and individuals…." Based on the assumption that people interact better with those with whom they share floors, the twenty faculty offices were divided among three floors, approximately seven per floor. The height of the building was limited to three stories so that people could use the stairs and meet in the stairwell, rather than being forced to use the elevator…. Corridors were to be wide enough to allow users to stand and talk comfortably. The central stair was to be spacious enough to encourage conversation….

WIDE CORRIDORS in the Lewis Thomas Lab work like a main street for each floor. Everybody encounters each other there, there's room for service vehicles, and there's space enough for some equipment—especially equipment that needs to be shared. Visible at the end of the corridor is one of the lounges. Every corridor has one at each end—they're used for small presentations and meetings, for hanging out, and for study.

1993 - PRINCETON'S LEWIS THOMAS MOLECULAR BIOLOGY LABORATORY (1986, right) inspired a twin, the George LaVie Schultz Laboratory (1993, left), which learned from experience with the earlier building. Facilities manager Anita Lewis-Antes says that among the successful ideas carried over to the new building were the numerous lounges, widespread use of daylighting, lots of interior wood finish, and easy access to commonly used equipment. *Not* repeated were exterior wood (too hard to maintain), reflective ceiling lights (overhead services shadowed them), and inflexible lab furniture (Schultz's cabinets and benches can be used either sitting or standing).

LABORATORY SPACE in the Lewis Thomas building is kept bright and cheerful by the abundant windows, but the services overhead prevented reflective lighting from working well.

Lounges with kitchenettes were planned as places for informal gatherings. Blackboards were provided in these lounges for impromptu work sessions. Offices were located where people would tend to cluster—in part to enhance competition by making it easy for users to see what others were working on…. The chairman wanted to invite undergraduates to mingle with faculty and postdoctoral researchers over tea in the afternoon; therefore 1,000 square feet were planned for a central meeting room…. A large lecture hall was planned to include a kitchen so that snacks could be served at evening lectures….

The architects and the scientists each drew up their own lists of required spaces. Jim Collins reconciled the lists and put together a two-page preliminary program summarizing square footage requirements. (Payette believes that designers do not read voluminous programs.)[2]

According to facilities manager Anita Lewis-Antes, the building is constantly visited and studied as a model research facility. She notes that the generous width of the corridors, designed for conviviality, turned out to be essential for materials-handling on flatbeds and glassware carts (that was a surprise not designed for), and the labs designed to share technical equipment—for conviviality as well as economy—were able to absorb many more users than expected when the building population grew from the anticipated 100 people to 300. Final proof of success: when the department outgrew the Lewis Thomas building, the same designers and ideas were employed for a nearly identical companion building 100 feet away, linked by a tunnel.

I should offer a counter-example, just to keep balance here. One of the most admired museum buildings in the world is the Monterey Bay Aquarium (1984) in California. "The building just evolved," says principal architect Charles M. Davis. No formal program. There, however, the architect had a very focused and design-savvy client in the person and family of electronics entrepreneur David Packard.

The great virtue of programming is that it deeply involves the users of a building and makes it really *their* building. The great vice of programming is that it over-responds to the immediate needs of the immediate users, leaving future users out of the picture, making the building all too optimal to the present and maladaptive for the future. An old saw of biology decrees, "The more adapted an organism to present conditions, the less adaptable it can be to unknown future conditions."[3] In practice, many a thoroughly programmed building is obsolete by the time it is built. Architect William Rawn has found a way to get around clients who become entranced with defining every detail of the building. He gets the client to discuss the *vision* of the building at the very beginning in some depth, and then he can harken back to that vision when obsessive details threaten to overwhelm the building with fussiness. Vision is generic, and generic is adaptive.

It's not that future considerations are left out of programming. Often they dominate the process. But it is narrow, wishful futures that are considered. "We'll wire the whole building with fiberoptic cable so we'll be ahead of the game when broadband technology comes on line." (Then office technology veers toward wireless instead.) "We only have funds for the core building now, but we'll leave the north wall windowless and made out of Dryvit instead of brick so it will be easy to add the north wing later." (The wing is never built, and the peeling, windowless wall depresses generations of occupants and passersby.)

The iron rule of planning is: whatever a client or an architect says will happen with a building, won't. Architects always want to control the future. So do clients. A big, physical building seems a perfect way to bind the course of future events. ("Once we move the company into the new building, then we can use it to limit our growth.") It never works. The future is no more controllable than it is predictable. The only reliable attitude to take toward the future is that it is profoundly, structurally, unavoidably perverse. The rest of the iron rule is: whatever you are ready for, doesn't happen; whatever you are unready for, does.

Programming cannot accommodate perversity. As Frank Duffy understates the problem, "A program ought to have in it the seeds of all subsequent change. It's jolly difficult." This is where scenarios shine, because it is their job to seek out and celebrate future perversity. Like programming, scenario planning is a future-oriented formal process of analysis and decision. Unlike programming, it reaches into the deeper future—typically five to twenty years—and instead of converging on a single path, its whole essence is divergence.

My familiarity with the process comes from six years' work with Global Business Network (GBN), which facilitates scenario planning for organizations ranging from multinational corporations such as Bell South and Hewlett-Packard to non-profits like the Sierra Club and the National Education Association. The technique we employ was developed to its current state at Royal Dutch/Shell, a Dutch-British oil company that rose from being the tenth largest corporation in the world in 1972, when it began using scenarios to guide strategy, to become number one by 1989, surpassing even General Motors and Exxon. The best book on scenario planning was written by my colleague at GBN, Peter Schwartz—a veteran of five years of scenario planning at Royal Dutch/Shell.[4] The scenario approach was having a vogue in American and European businesses by the early 1990s, but it was overlooked for building projects, where I think it is ideally suited.

Scenario methodology is pretty easy to summarize. I'll describe how it works on the scale of a corporation or major building, and you can imagine how a quick-and-informal version might serve, say, the designing of a summer cottage.

---

[2] Ellen Shoshkes, *The Design Process* (New York: Whitney, 1989), pp. 104-105. See Recommended Bibliography.

[3] It is in "unknown future conditions" that opportunistic, "r-selected," Low Roadish species most prosper. Weeds love change.

[4] Peter Schwartz, *The Art of the Long View* (New York: Doubleday, 1991). See Recommended Bibliography.

Whoever is going to lead the scenario exercise begins by interviewing the major players in the organization or the project at hand, to pick up their vocabulary, the major issues ("What keeps you up at night?"), and the consensus expectations about the future. Then the policy-making people and a few of their advisors are gathered for a two-day session. The first day begins with identifying the focal issue or impending decision that makes the scenario exercise necessary. ("We're growing to death! Should we stop growing and consolidate? Or find new directions where growth will more than pay its way?")

Then the group explores the "driving forces" that will be shaping the future environment. In business, driving forces often include changes in technology, regulation, the competition, and the customers. For a building, driving forces might include changes in technology, in the neighborhood, in the economy, and in tenant use. The group ranks these driving forces in terms of importance and uncertainty, placing the most important and most uncertain highest, because it is the important uncertainties that will drive the scenarios apart. At the same time the group should be identifying "predetermined elements"—reliable certainties such as the aging of baby boomers—which will be in all the scenarios.

Now, working with the crucial uncertainties, the group identifies scenario *logics*—basic plot lines. ("If we have rapidly changing communication technology *and* a boom economy, this office building is going to face constant services refits and a high turnover in tenants.") The goal is to develop scenarios that are both plausible and surprising—shocking, in fact. One artful way to do that is to identify and spell out the "official future"—the future that everyone thinks they are supposed to expect. Make it one scenario. ("We'll move into the cottage permanently when we retire.")

Then start thinking the unthinkable. Let people top each other in imagining terrible and delightful things that might happen, exacerbated by the crucial uncertainties. ("Both of us are laid off, and can't get work, and we have to sell the house and move into

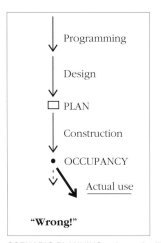
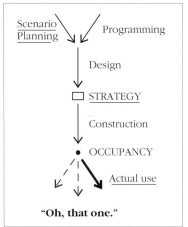

SCENARIO PLANNING reduces the likelihood of being pushed around by a building obdurately clinging to a future that never happened. It reduces surprise in a good way. When something untoward happens, the building is ready for it.

the cottage next year! Or we keep the house and have to sell the cottage next year!") Typically these scenarios soon acquire a frightening plausibility which makes a mockery of the "official future." Dutch scenarioist Kees van der Heijden says, "Good scenarios always introduce an element of novelty in an area of anxiety—a new concept, thought, or discovery."

Sleep on that. Next day the group revisits the preliminary scenarios, adjusts them—often radically—and then begins to flesh them out into detailed, vivid stories. Naming the scenarios is important. The names should have an overstated, caricature quality—"It Came From Outer Space"; "Retailers Win"; "Litigation Hell"—since they will be used as shorthand by the planners. There should be two to five scenarios—no more. The *probability* of one or another happening is not useful to explore, except for including one or more "wildcard" scenarios considered to be quite unlikely but horrifying if they occur. ("Our core business is obsolesced overnight, and we have to rebuild around one of our fringe activities.")

Now the group goes back to the focusing issue or decision to devise a strategy that will accommodate all the scenarios. The thing to avoid is a "bet-the-company" strategy that wins in only one scenario and loses in all the others. One approach is to develop a "robust" strategy that is viable in the variety of futures. Sometimes it is discovered by "regret analysis." You ask, "What if we get it wrong? What would we regret not having done? What would we regret locking in?" You're seeking to balance your risks intelligently. Another approach is devising an "adaptive" strategy that is exceptionally alert to changing events and can adjust quickly.

As the new strategy emerges, it changes how the scenarios would play out, so the group needs to cycle through the process a few times to get a set of scenarios and a strategy that make sense with each other. ("Okay, it's a smaller, more finished cottage that can be easily sold, or lived in immediately, or grown gradually over time.") The final task is to identify some leading indicators that will be monitored to see which scenario (if any) is actually occurring in life. Then the group must return from the mountain and communicate the strategy *and* the scenarios to everyone in the organization or working on the project, because subsets of the overall strategy need to be devised at every level.

When and how would scenario planning fit into a standard building project? Two points in the usual design sequence come to mind. Sometimes both might be used, since scenarios frequently are developed by stages, with research in between to pursue the new questions that scenarios inevitably raise. The first opportunity would be the initial forming of the vision for the building. The second would be after there has been some preliminary programming, which might be used as the "official future" of the building, and which a fresh look at scenarios should challenge.

Participants in the scenario exercise might be different than in programming. It should include policymakers and stakeholders and senior people involved with design—those responsible for the long-term purpose of the building, the building as investment, and the shaping of the building. For a commercial building, the initial scenario session might include the architect, the head of the organization planning to use the building, the organization's project manager, the main tenant, and the property owner. A follow-up scenario session would have the same people plus the project engineer, the facilities manager, someone from the bank that holds the mortgage, perhaps a friendly neighbor, perhaps someone from the city's design review board or even city council. (For some design firms this use of scenarios would be a variation on what they already do with "design charettes.")

A building is a huge investment, a black tarry pit of sunk costs, a trap and a prison. The job of scenario planning is to question whether a building is really needed at all and, if it is, to convert it from a potential prison into a flexible tool. Veterans of bad decisions are full of warnings: "If you ask an architect to solve your problems, all you're going to get is a building." "Building a house can destroy a troubled family." "Never expect a building to solve organizational problems." "To an experienced board member, when you see a business getting a big new building, it's a bad sign."[5] "Our solution should have been a real-estate strategy rather than a design decision." The trick is to frame the building question in a larger context, so instead of leaping to "We need a bigger building," really explore the more general problem, "We need to handle our growth."

A film and television production company in San Francisco, modestly named Colossal Pictures, was so beset with growing pains that it had to consider completely renovating and expanding the former warehouses it was using for studios and offices. Working with architects Richard Fernau and Laura Hartman, and with Lawrence Wilkinson and myself from Global Business Network, the company undertook a preliminary scenario exercise. Two major driving forces were seen as influencing the company's growth. The industry might settle into a more stable market, or it

---

5 Lloyd's of London again comes to mind. Its monumentally expensive headquarters was completed in 1985. From 1988 to 1993 the company lost billions of pounds each year, suffered from severe capital shortage, and got into deep trouble. Judicious scenario planning might have headed off the extravagant building project.

might become even more highly fluctuating. And the company might keep its competitive advantage by remaining small and agile, or it might gain economies of scale from growing big and powerful.

Laid out in a matrix, that gave four potential scenarios. The combination of small size in a stable market was named "Boutique." Small size in a fluctuating market was "Art Commandos." Large size in a stable market was "800-Pound Gorilla." Large size in a fluctuating market was called "Spanish Armada," and was regarded as a danger to be assiduously avoided; furthermore, an overspecialized building could easily add to the danger. The company decided that its strength (and its delight) lay in being Art Commandos, but it needed to be able to survive in either the Boutique or Gorilla scenarios. Plans for gaudy new facilities were shelved, and the main existing building was re-examined in terms of being able to expand into adjoining space or to sublet, if necessary, some of the current space. Revamping of the existing space was explored strictly in terms of relatively inexpensive changes that would make project coordination easier. Schemes for elaborate tailoring of the building around existing or expected special-effects technologies were dropped.

Architects are ideally suited to lead scenario exercises. They are imaginative and enjoy spinning fantasies. Often they are eloquent storytellers. Their enormous graphic skills let them sketch alternative ideas quickly and realistically. Many are able team leaders, good at pulling ideas out of a group and then integrating them. They are used to grasping and shaping large, complex schemes. And they are accustomed to exploring alternative design paths (though usually out of sight of clients). Fernau and Hartman broke that custom in their work with Colossal by providing an early model of the building which was completely demountable. Pieces could be moved around to demonstrate several different ideas the architects were pursuing, and anyone in the planning group could reach in and try out a different idea.

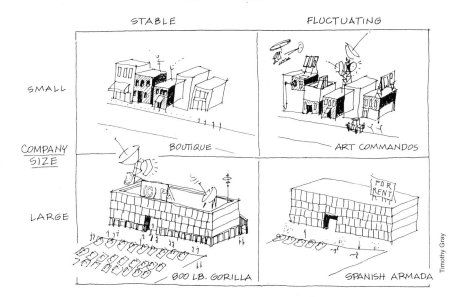

SCENARIOS often emerge from a matrix diagram showing various combinations of driving forces (here, market fluctuation) and/or organization strategies (here, company size). In these drawings by Timothy Gray, who worked on the project for Colossal Pictures, the "(C)P" is the company logo. Strictly for enhancing discussion, the drawings represent caricatured building implications rather than actual plans.

A ROUGH, DEMOUNTABLE MODEL of Colossal's warehouse (22,000 square feet) was provided by Fernau & Hartman early in the project and was always in the middle of the table during discussions with the client team. Anyone could reach in and move things around.

PROSPECTIVE BUILDING ELEMENTS were crafted separately to be shifted about in the Colossal model. Some here include stairs and walkways, entrances, and project meeting rooms (called "war rooms" by the client).

Architects weary of dealing with fragmented clients are likely to welcome one of the major advantages of scenario planning, which is to provide a shared language and frame of reference among most of the decision makers. Since it invites and exploits divergent views, scenario planning can even bridge conflicting factions. The process has been a major tool for ending apartheid in South Africa, because both sides could participate without giving up their own experience and identity. It is seen there as "a tool for collaborative strategizing in politically charged public debates."[6]

Scenarios can provide a way around some of the major design errors that commonly occur with buildings. The worst mistakes come not from wrong decisions but from not doing the right thing *that never occurred to anybody to do*. Scenarios attack a design from so many directions that gaps and oversights are likely to show up. Also, scenarios severely test fond notions that might otherwise get by unchallenged. ("We will have the stairs run clear around the periphery of the atrium like a magnificent spiral staircase.") Grandiose ideas that don't work are more hated by building occupants than almost anything else. Scenarios are conservative—they keep you from betting the whole building on some potential folly—but they are also innovative. Ingenious design ideas can emerge from the creative process of scenario-crafting and strategizing. ("Both grandkids said they love playing in low-ceilinged rooms, so we'll save money and bulk with an old-fashioned half-height cottage attic, and they can have a kid heaven up there.")

"Our most important responsibility to the future is not to coerce it but to attend to it," observes city theorist Kevin Lynch. "Collectively, [such actions] might be called 'future preservation,' just as an analogous activity carried out in the present is called historical preservation."[7] "Future preservation" means that the building is not only built to last, but it remains always capable of offering new options for its use. Freedom to adjust and even to change direction entirely is preserved.

A number of professional tools are already available that would assist strategic thinking with buildings. One is life-cycle costing, which can help raise the thinking of decision makers out of the immediate shadow of the building's mortgage. Careful consideration of long-term costs of operation, maintenance, repair and remodeling, services refits, and eventual demolition and disposal can inform the various scenarios and guide design decisions toward strategic scope. Spend a little more now (say, on roofing and insulation) to save a lot more later (in maintenance and energy costs). Build something smaller and more solid now that can expand in a variety of directions later; or, contrarily, build something that seems too large now but can be grown into easily.

Another tool coming to hand is "virtual reality"—computer-displayed models of buildings which people can bodily "enter" and walk around in. One of the pioneers of the technology, Frederick Brooks of the University of North Carolina, describes how virtual reality has worked for building projects there:

> You *have* to iterate. What people do with virtual buildings, if they're willing to take the time, is they take the floor plans and they live through a day and trace all the patterns and see where they run into trouble, where they store things, and where they have traffic problems, and so forth. And they run through the major annual events and whatnot. The architects say they don't have any trouble visualizing three-dimensional buildings from floor plans, but the *clients* surely do. So, if the client and the architect working together can debug the

6 An accessible example of scenarios at work in South Africa is *The Mont Fleur Scenarios*—which took place in 1991 and 1992, with 22 participants on the scenario team from the African National Congress, Pan-African Congress, government, business, academia (notably economists), and trade unions. After a sequence of meetings, the four scenarios that emerged were called: "Ostrich" (non-representative government); "Lame Duck" (long transition); "Icarus" (fly now, crash later); and "Flight of the Flamingoes" (inclusive democracy and growth). The results and process of the scenarios were publicized in South Africa newspapers, and a 30-minute video was widely distributed (available from The Institute for Social Development, University of the Western Cape, Private Bag X17, 7535 Bellville, South Africa).

7 Kevin Lynch, *What Time Is This Place?* (Cambridge: MIT, 1972), p. 115. See Recommended Bibliography.

building at what is called the design development stage, while everything is still on paper and you haven't even done the working drawings, you can save a lot of trouble downstream.[8]

Most design in the real world is guided by rules of thumb. What might be the rules of thumb for strategic building designers? Some can be borrowed directly from chess players: "Favor moves that increase options; shy from moves that end well but require cutting off choices; work from strong positions that have many adjoining strong positions."[9] More specific to buildings: overbuild Structure so that heavier floor loads or extra stories can be handled later; provide excess Services capacity; go for oversize ("loose fit") rather than undersize. Separate high- and low-volatility areas and design them differently. Work with shapes and materials that can grow easily, both interior and exterior. "Use materials from near at hand," advises Massachusetts builder John Abrams. "They'll be easier to match or replace."

A spatially diverse building is easier to make use adjustments in than a spatially monotonous one—people can just move around. Medium-small rooms accommodate the widest range of uses.[10]

When in doubt, add storage. Add nearby storage—closets, cabinets, shelving; and deep storage—attics, basements, unfinished rooms without windows. What begins as storage can always become something else, and if it doesn't, there's never enough storage anyway.

Shun designing tightly around anticipated technology. As energy analyst John Holdren says to all futurists, "We overestimate technology in the short run and underestimate it in the long run." So design loose and generic around high tech. You will be wrong about what is coming, and whatever does come will soon change anyway.

Playing out in imagination the potential consequences of even minor-seeming details is worth doing. ("We could put a tall window with a low sill here so it can be changed easily to a door later when we add the deck. But what if the deck turns out to be a porch and it winds up somewhere else because of a tree or something, and a drunk guest walks through the tall window one wintry night?") Scenario-makers love imagining disasters. They are right to do so.

Remodeler Jamie Wolf is convinced that it is lack of imagination that leads to such incurable problems as narrow halls in apartments: "Who was the decision maker who gave us that hallway and engendered the curses of a thousand tenants? Usually it is dimension by default. 'Hmm, what does the building code allow as a minimum; okay, I'll make it that wide.' Or, 'If I can take six inches from the hallway and six inches from the closet, I'll save some money over here.' The decision is made by criteria that never imagine the constraint placed on every inhabitant of that building for the rest of its life."

"Loose fit" provides room to adjust, but what do you do about room to grow? Perfect fit is fleeting. Yet organizations are always striving to occupy just one building, both for convenience and to avoid divisiveness. The Digital Equipment Corporation was inspired by its founders' experience in funky Building 20 at MIT to start its business in one corner of a huge abandoned fabric mill in Maynard, Massachusetts. As its prosperity grew, it expanded into the empty space. That experience of physical freedom became a core part of the corporate identity and culture. For decades, though it had spread through scores of buildings, the company

---

[8] Quoted in Howard Rheingold, *Virtual Reality* (New York: Simon & Schuster, 1991), p. 42.

[9] Kevin Kelly, *Out of Control* (New York: Addison-Wesley, 1994).

[10] "Studying the history of British hospitals... Peter Cowan found that rooms ranging in size from 120 to 150 square feet were those that most easily converted to a wide range of new uses. Smaller rooms tended to resist conversion except for a few new uses, and the number of new activities for which larger ones were reusable did not increase much despite the increase in size." Kevin Lynch, *What Time Is This Place?* (Cambridge: MIT, 1972), p. 108.

PAVE WHERE THE PATH IS. An oft-told story (perhaps apocryphal) tells how a brilliantly lazy college planner built a new campus with no paths built in at all. She waited for the first winter and photographed where people made paths in the snow between the buildings. Next spring, that's where the paving went. Some design is better if it's postponed.

1990 - Real traffic went one way while the planned path went another at the University of Missouri. People always take shortcuts. Why not do it their way? This shortcut has a sweet curve to it, missing in the boring planned path.

May 1990. Brand.

1991 - More oppressive is MIT's treatment of a pedestrian shortcut. Ugly yellow rope is strung to prevent people from taking the natural route straight ahead to the corner of the building. It's a major thoroughfare for people on foot, but everyone has to detour clear around to the right—50 yards out of their way. The greensward so diligently protected is unused and unloved. With the rope around it, it's just a stupid barrier.

2 May 1991. Brand.

kept a few spaces in the original mill sublet to other tenants as a reminder of the need to retain the freedom to shrink as well as to grow.

The most convenient form of expansion is cellular. As my wife discovered with her mail-order business, fast-growing, high-risk entities such as new companies do best leasing space in long, old, cheap buildings made up of a series of bays divided by simple walls. If the company expands, wait until a neighbor start-up company fails or moves away (which they do frequently) and take over its space, knocking a few doors through the wall. In hard times, shrink back and seal up the wall. This is the common dynamic of commercial main streets, where successful shops expand into adjoining buildings. The same flexibility is found in the vertical dimension with old townhouses, which subdivide into flats and then recombine again as the real-estate toniness of the neighborhood scales up and down.

Taking a strategic approach to a building may mean postponing many design decisions and leaving them to the eventual users of the building. This is heresy to professional designers who want all the design problems solved beautifully once and for all. It was developers, not architects, who recognized the market for "shell-and-core" office buildings, where the tenants instead of the builders fit out each floor. But there are potential advantages to architects.

In the case of Colossal in San Francisco, the use of scenarios

inspired an epiphany partway through the design process. The client realized there wasn't money or time to completely remodel the building, but it could be taken by stages, if Fernau and Hartman would agree to working that way. The architects seized the idea. It would give them reliable ongoing business (treasured by architects), and it challenged them to develop an evolutionary design. They came up with a scheme which specified some areas in the building to be "cooked"—highly finished and flashy—and some areas left "raw"—unfinished but usable. The raw parts could be finished later in response to accumulating revenue and to all that would be learned from use of the cooked parts. In the meantime the raw areas could be tailored roughly by whoever was in them.

Postponing some of the design yields more building for less money, since less is spent on detail design and detail finish. And adaptivity is built in.

188

1782 - Design for a Metropolitan Cathedral, by the Paris architect Etienne-Louis Boulée (1728-1799). He had a major influence on the architecture of the Napoleonic period.

TWO VISIONS OF ARCHITECTURE. Boulée (left) wanted people awed, tiny, and powerless before the magnificence of the architect's achievement. The mother on the right has a different attitude about her creation.

A building is not something you finish. A building is something you start.

"Evolutionary design" sounds like a contradiction, but it needn't be. In the 1980s, both ecology and economics underwent a quiet revolution when they began to realize that natural and market systems were "variance-driven" rather than "equilibrium-based." There is no "climax" in ecological communities; irregular oscillation drives the continued self-tuning of the system. Adaptivity was said to thrive "on the edge of chaos." Likewise in economics. With the failures of the world's command economies (communist and strict socialist), it was noted that governments that institutionalized stability suffered systemic chaos, whereas market economies and democracies that institutionalized chaos enjoyed systemic stability. Command economies collapsed. Market economies muddled through. By making more mistakes they had less failure.

The Darwinian mechanism of vary-and-select, vary-and-select has one enormous difference from the process of design. It operates by hindsight rather than foresight. Evolution is always away from known problems rather than toward imagined goals. It doesn't seek to maximize theoretical fitness; it minimizes experienced unfitness. Hindsight is *better* than foresight. That's why evolutionary forms such as vernacular building types always work

better than visionary designs such as geodesic domes. They grow from experience rather than from somebody's forehead.

What does that leave for "evolutionary design"? Sometimes you have to shape a new kind of building and you don't have a century for generations of cut-and-try. One strategy, which scenarios encourage, is to make the building responsive to *future hindsight*—perpetual later reappraisal and adjustment. Focus on structure rather than details; leave parts of the building uncooked; use materials and forms that are inexpensive to change around.

One way to institutionalize wholesome chaos is to disperse significant design power to the individual users of a building while they're using the place. Notice the difference between kitchens designed to be used by powerless servants—they are usually dark, cramped pits—and kitchens used by the heads of a family—bright, spacious, centrally located, crammed with conveniences. A building "learns" only through people learning, and individuals typically learn much faster than whole organizations. This suggests a "bottom-up" rather than "top-down" approach in the building's human hierarchy. Robotics researchers call it "subsumption architecture"—pushing the power

"Water Baby." Photographer/Director: Mike Portelly. From a TV commercial for: British Gas PLC.

comes naturally because they have a hands-on relationship with their space, and they know how it actually works and will have ideas about how to improve it. Some will set about hot-rodding the building, and the rest will appreciate what they're up to.

The architectural profession has emasculated itself far enough. What is suggested here is a new dimension—architects maturing from being just artists of space to artists of time. Building aesthetics might acquire a new edge. Artist/musician Brian Eno observes:

> It seems to me that the best designs are those which accommodate the most contradictions. Looked at the other way, the most boring design is that which is directed at a simple, well-defined future. A lot of New Age music exemplifies this, as does, for me, Le Corbusier. They are both addressed to simple world pictures, and to simple ideas about how humans behave and what they want.

It wouldn't take much adjustment to unleash the full ingenuity of architects on the juicy problems of designing for time. They could supplement the dutiful process of programming with the enjoyable practice of scenario planning. They could do more post-occupancy evaluation, particularly of their own buildings, but also of existing buildings that relate to new projects. They could seek the stability of ongoing relationships with clients instead of the all-at-once, do-or-die, design-crisis approach now employed. They could seek new ways to employ time as a tool in building design and use. "Time," wrote Francis Bacon in 1625, "is the greatest innovator."

to respond to the bottom of the organization. Their motto is "Fast, cheap, and out of control!" Perception and responsive action take place locally rather than being mediated through some remote command center. There's no need to completely redo obnoxious overhead fluorescent lights when people are free to install their own preferred task lighting in their work areas.

What devices best encourage distributed "learning" in a building? Most are probably managerial. Let people try things. When some of them don't work, honor the effort, notice the lesson if any, and move on. (And stifle the deadening grumble of oldtimers, "We tried that; it didn't work!") Sometimes it is worth letting different groups try different variations of something, and race them. Then go with the winner. If one section of the organization proves to be a particularly effective innovator, give them even more latitude. One company I know of has the facilities manager's office try out every new idea that comes along. Encouraging variety in the organization gives everybody a richer mix of ideas to shop among.

What would a building look like and act like if it was designed for easy servicing by the users themselves? Once people are comfortable doing their own maintenance and repair, reshaping

CHAPTER 12

Built for Change

THIS BOOK has been assembling what might be called "steps toward an adaptive architecture." Has it arrived anywhere? What have we really got so far? If we were going to construct an adaptive building, what would be different about it? On the surface, not much, probably. An adaptive building would not surprise us if it started conventional and became unique, started conservative and became radical by being true to its unique life.[1]

Initial conservatism would be the natural result of a scenario approach to design, with its cautiousness about innovation and its imperative to protect the option of varying paths of development for the building. Furthermore, anything borrowed from local vernacular design is bound to look conservative—for the very reason that it is the familiar, well-tried, nuanced product of local climate, local culture, and standard building use. Also traditional materials, which age well and take advantage of deep experience in the building trades (and avoid the chanciness of trendy new materials) can be counted on to look good but ordinary. And respect for existing older buildings nearby obliges a polite blending in by anything new. Honoring the future begins with honoring the past.

Time-savvy conservatism would affect all of a building's basics—its design process, budget, technology, overall shape, space plan, materials, and structure.

The shaping of a confident, careful building cannot emerge from fragmented design. Design decisions will continue to be made right through the whole sequence of permissions, site preparation, construction, finishing, and inhabitation. That's not the way it's supposed to happen, but it's the reality. Revisiting and changing design decisions must not be allowed to stop or to confuse the drive toward completion. The client population should be able to speak through one person who reflects their experience and desires and has the authority to make decisions in their behalf.[2] The developer, project manager, or architect needs a similar authority over the design and construction professionals.

The major difference in a "learning" building is its budget. Following Chris Alexander's formula, there needs to be more money than usual spent on the basic Structure, less on finishing, and more on perpetual adjustment and maintenance. The less-on-finishing part takes forceful management because the architect will want pizzazz in the finish, where it shows, and many artisans are not happy skimping on finish quality. The architect needs financial incentive to hold the line. I have suggested settling on a preliminary design and construction budget, with the architect getting a flat fee—plus a generous bonus if the budget and schedule are met. This encourages realism and rewards parsimony.

If you want a building to learn, you have to pay its tuition. Maintenance alone runs $2 to $5 per square foot per year in commercial and institutional spaces, according to facilities manager folklore. Since about 30 percent of the operating costs in most buildings goes to paying for energy, significant money for maintenance, tuning, and remodeling of the building can be freed up by designing in energy efficiency through well-proven techniques—insulation, tightly crafted windows and doors, orientation to the sun, use of foliage (for summer shade), and appropriate color (light in hot climates, dark in cold).

---

[1] Progressive high-style buildings take the opposite course. They start radical and then are forced to become conventional in order to work at all well. This is another reason many architects hate to revisit old projects.

[2] New buildings, thanks to their enormous expense and apparent permanence, inspire ferocious anticipatory turf combat. Some architects refuse to design houses for argumentative families for the same reason that police hate to be called to break up domestic disputes—it's too violent. Clear authority is the solution. It doesn't matter whether Mom or Dad is the conduit of decision with the architect and the contractor so long as only one person is. If the client is an organization, the more senior the decision maker the better. If there is a project director, that person should report directly to the top.

1986 - The traditional northern European three-aisled structure lives on in US Army barracks that were built by the thousands during World War II. Each of the two floors was one long open space, with sergeant's quarters at one end and a latrine at the other, and a row of columns marching down each side, with bunks lined up outside them. This building—at Fort Cronkhite, near San Francisco—was taken over by the National Park Service. In the 1980s this upper floor was remodeled as an office for the Golden Gate Energy Center, with the columns anchoring the new walls and partitions.

COLUMNS PROVIDE A PHYSICAL GRID for space plan changes. They make it easy to imagine changes, easy to put them in, easy to remove them. They also are handy as risers to run cable, conduit, and other services.

1993 - In 1987 the space was taken over by the Pacific Environment & Resources Center. Seeking better privacy in the offices, in 1990 they grew the partitions on the left to full walls, and added walls on the right and in the foreground. The internal windows were recycled from other military buildings being dismantled by the Park Service.

1993 - The wide overhanging roofs (even on the gable ends) of the barracks shielded the sash windows from direct sunlight and also protected the walls from rain and sun damage. Fifty years after their hasty construction, the buildings are in fine shape.

A more radical monetary strategy is to attack the whole idea of a mortgage and its constant bleeding of money away from building upkeep. Sixty percent of the final cost of a mortgaged building disappears as interest to the bank instead of going into the building. What if the usual down-payment money were spent instead to complete a small core building or a large but rudimentary building? Then the usual interest money, instead of going away to the bank, would go into prolonged, attentive growth and improvement of the original building. Contractor Matisse Enzer, who promotes this idea, estimates he could build a minimal but livable "continual house" for the standard down payment and then be paid the usual rate of interest to grow the house bit by bit over

the following years. "With ten or twelve projects like that going on, I could make a good living." The owner gives up some tax advantages but gains freedom from bondage to the bank and eventually gets more house—and a far more highly adapted house—for the same money as a mortgaged monster.

The temptations of being in fashion are similar to the mortgage temptation—buy too much now, pay too much later. Another reason for conservatism in design is to avoid handicapping the building with passing fancies of the moment, whether of functional layout or of style. Any too-contemporary style loses value from the instant it is built, and often it defies good sense, since it is untested. The A-frame vacation houses of the 1960s hemorrhaged energy through the huge windows, and the lack of ceilings made them impossible to heat. They wasted space, and the pitched roof-walls were unusable for anything, but everybody built the things.

How does one shake free of the aesthetics of the moment—real-estate fads and the vogues of magazine architecture? Maybe scenarios can help. If you design a building that you think tourists would admire and envy in ten years, and that preservationists will fight to save in fifty years, you'll probably get the proper mix of bemused conservatism and mythic depth. Freed of fashion, a building can become honestly interesting in its own terms.

The temptation to customize a building around a new technology is always enormous, and it is nearly always unnecessary. Technology is relatively lightweight and flexible—more so every decade. Let the technology adapt to the building rather than vice versa, and then you're not pushed around when the *next* technology comes along.

As for shape: be square. The only configuration of space that grows well and subdivides well and is really efficient to use is the rectangle. Architects groan with boredom at the thought, but that's tough. If you start boxy and simple, outside and in, then you can let complications develop with time, responsive to use. Prematurely convoluted surfaces are expensive to build, a nuisance to maintain, and hard to change.

The way rooms and floors are laid out can be crucial for a building's resilience to changing times, but little research has been done on the subject. The best I've seen is in Anne Vernez Moudon's exhaustive study of San Francisco's famous turn-of-the-century "painted lady" Victorian row houses. Titled *Built for Change*, the book analyzes the elements that made these spec-built wood-frame confections such survivors. The small, narrow lots—25 to 30 feet wide, 100 to 137 feet long—made for dense, convivial neighborhoods. Behind the appealing variety of house fronts was an absolutely generic space plan. Stairs and corridors

From *Victorian Classics of San Francisco* (Sausalito: Windgate, 1987), plate 37. A reprint of *Artistic Homes of California*, 1888.

1887 - An early prototype for the hundreds of Victorian semi-detached row houses in San Francisco was this residence at 2202 California Street. The tall brick chimneys would not survive the earthquake of 1906, and the elaborate Queen Anne facade detailing evidently required more maintenance than was available. Even the beautiful leaded windows went the way of all windows. But the house itself had the resilience to survive intact.

SAN FRANCISCO'S VICTORIANS, long scorned for being over-ornate, survived to become one of the city's signature attractions. Though wood-frame, and their ornament invited rot, they survived because their room layout was so adaptable.

were on one long side, a row of rooms off the corridor on the other. Moudon:

The hall averages six feet in width for the entire depth of the box. Because its dimensions are generous, it can accommodate functions beyond circulation. Bathrooms can be squeezed in, leaving a three-foot passage, adequate for walking. Closets can be placed wherever leftover space occurs, easily leaving some three feet of hall space. The interior stairs are a standard item always inserted in the front of the house near the entrance, accentuating the grandeur of the hall. The rear of the hall, which can have prime access to light, can then form a single small room. Thus the hall, which diagramatically appears like mere circulation space,

1991 - As if worn smooth by time, the house has lost its ornament, and the roofs of the turret and entry have been simplified. A dormer and additional window have shown up next to the turret, the former basement now appears to be lived in (was the whole thing jacked up?), and the entry has moved down to its level. Fire escapes festoon the building.

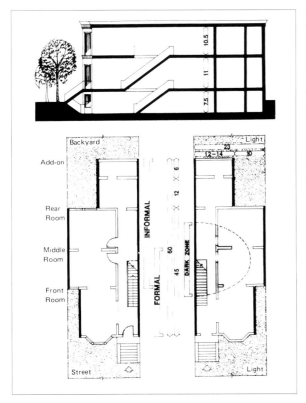

THE VICTORIAN BOX that stands behind the boggling variety of San Francisco row house facades is laid out in one universal pattern. The wide hall is on one side, the many rooms on the other, with recesses to allow daylight into the middle rooms. The rooms offer the option of opening into one another in various combinations, or of opening only into the hallway. In most houses the combinations change as often as the tenants do. This illustration is from Anne Vernez Moudon's *Built for Change*, p. 57. (See footnote below.) The measurements are in feet.

becomes the support core that relieves the rest of the box of clutter. The generous width of the hall is also the major reason for the inherent flexibility of the box.[3]

The three to six innocuous-looking rooms lined up on each floor were enough to make the Victorians later divisible into two or three flats, one per floor. The rooms are modest in size (averaging twelve by twelve feet) and unspecialized in function. Each is lit and ventilated with its own windows, each has access to the corridor, each is capable of opening into adjoining rooms. Observing the boundless flexibility of this arrangement, Moudon draws a radical conclusion: "A return to the room as module for residential design is a necessary step toward creating resilient space. We must abandon the use of *dwellings* as modules of spatial organization."[4] The problem with dwelling-as-module (as the fundamental design unit) is that it encourages overspecialized rooms that will be difficult or impossible for later tenants to respecialize for their own needs, whereas simple, autonomous rooms can be constantly readapted without stress to the building.

If you want a lovable building, a strategic decision needs to be made right at the beginning. The design and construction can fruitfully take either the High Road or Low Road, toward beloved permanence or toward beloved disposability. The High Road requires Structure built to last and some areas of very high finish indeed, particularly with the Skin and at least some interiors, to set a high standard for future work. The major threat to an urban High Road building over time is shifting real-estate values, so either a financial endowment or great public esteem is needed to protect the property.

---

[3] Anne Vernez Moudon, *Built for Change* (Cambridge: MIT, 1986), p. 65. See Recommended Bibliography. I asked Moudon if she thought the San Francisco Victorians were designed for change. "No," she said, "they were just designed to appeal to a variety of tenants."

[4] Title above, p. 188.

A Low Road building needs only to be roomy and cheap. Structurally it should be robust enough to take the major changes in use it will attract. Finish can be minimal and ornament modest or absent entirely. Initial Services can be rudimentary. Design it primarily for storage and it will soon attract creative human occupants.

Longevity has no chance without serious Structure. "A building's foundation and frame should be capable of living 300 years," says Chris Alexander. "That's beyond the economic lifetime of any of the players. But construction for long life is what invites the long-term tampering it takes for a building to reach an adapted state." The lack of economic incentive suggests a role for government, using building codes, tax credits, and even direct sponsorship to get buildings that will serve the community for generations. Some of the solidest buildings in America were constructed during the 1930s by the New Deal programs of the Public Works Administration and the Works Projects Agency—libraries, schools, park buildings, bridges, dams, viaducts. That investment by government in "infrastructure" is seen by historians as having been the basis for the economic boom of the 1950s and 1960s. America is said to be overdue for a new round of infrastructure investment.

Thanks to growing environmentalism, another long-term issue is conserving the "embodied energy" of buildings and cutting down on the enormous solid-waste burden of demolished buildings.[5] Recyclable building materials are the obvious solution. Some industries, led by the German auto manufacturers, are adopting "design for reuse" (DFR) and "design for disassembly" (DFD) engineering. Design for disassembly in building construction is doubly appealing because it invites later reshaping of a building even at the Structural level. The present wasteful design-for-

demolition practices in the building trades are facing a revolution that will change the behavior of everybody from the materials manufacturer to the architect to the carpenter. Nailing light-construction buildings together, for example, is archaic. It's as easy to power-drive (and undrive) self-tapping screws, and the cost of lumber keeps going up, making wood valuable enough to recycle.

Wood is already the most adaptive of all building materials because amateurs are comfortable messing with it. Easy disassembly would help even more. When stud walls were first invented in Chicago around 1833, the technique was call "balloon frame" derisively because it looked so weak and ephemeral. It wasn't weak at all, but it was relatively ephemeral, certainly when compared to the old timber frame buildings with their massive posts, beams, and rafters fastened together with clever joints and wood pegs. Timber frame was the original design-for-disassembly building material—just knock out the pegs. Throughout medieval northern Europe, timbers were handed on from building to building for centuries. Recycled timbers were used even in

National Archives. Neg. no. 135-SA-01319B

1934 - A PUBLIC WORKS ADMINISTRATION LIBRARY in Allenstown, New Hampshire, was one of thousands of durable buildings built by the government in the 1930s. With its slate roof, brick walls, and simple plan, it was the embodiment of taking responsibility for the long term. Inside, there was a lobby in the middle, a children's reading room on the left, an identically-sized adult's reading room on the right (both with fireplaces), and a private work area added on to the back. The design of the PWA libraries was based on the pattern of the 1,679 free libraries built by Andrew Carnegie between 1886 and 1917.

---

5   The second-largest portion of landfill waste disposal is taken up by building debris, according to definitive studies reported in William Rathje and Cullen Murphy, *Rubbish!* (New York: HarperCollins, 1992). The largest portion is paper.

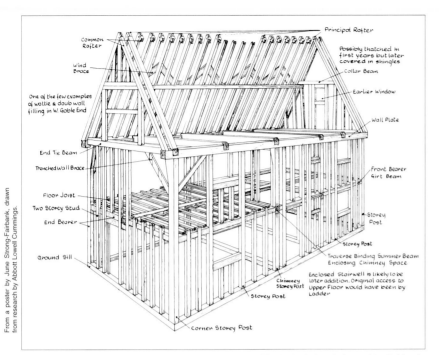

From a poster by June Strong-Fairbank, drawn from research by Abbott Lowell Cummings.

TIMBER-FRAME HOUSES are multi-generation in longevity. This is the timber skeleton of the Fairbanks House in Dedham, Massachusetts (see p. 121). Built in 1636, it is the oldest wood-frame building in America. After three and a half centuries, it is nearly as strong as ever.

1973 - The Stonor House in Oxfordshire dates from circa 1300. Over the centuries, interior walls, floors, and rooms like this top-floor bedchamber come and go amid the massive structural timbers.

21 February 1973. Royal Commission on the Historical Monuments of England. Neg. no. BB73/1700.

prestigious structures for royalty—a fact which has given building historians fits: dating a building accurately becomes nearly impossible when all the materials come from different eras. The current timber-frame revival is worth encouraging. (Stick with traditional hardwood pegs, incidentally; they last longer than metal bolts.) Timber frame assures a 300-year Structure because of its massiveness and because the building's endoskeleton is well protected from the weather outside and exposed for inspection inside. It is adaptable, recyclable, and beautiful.

The simpler a roof—pitched, of course—the less it leaks or needs maintenance. Complexity can come later if it must. Roofs that are built fussy at the beginning are an obstacle to later change. The more a roof overhangs, the better it protects the walls from sun and rain. The lighter the color of the roof, the better it will stand up to solar deterioration, keep the building cool, and lessen the stress of temperature change in the roof materials.

Walls, especially if they are separate from Structure, can be taken more casually. They probably should be, since they are the part of a building most visible and most confining to people and therefore most likely to change. In order to accommodate growth in any direction, they should be vertical and should begin flat and simple, like a good initial roof. They should invite easy penetration with new doors or windows. This is the great attraction of stud walls.

I see no reason for most American housing still being made of wood 2 x 4 studs. Nearly all new European houses, most Japanese houses, and nearly all American commercial and institutional interior walls are made with steel studs. Galvanized steel studs are cheaper, lighter (one-fourth the weight of wood), straighter (no warp), easier to cut (metal snips), and conveniently holed for stringing conduit and pipe. They don't burn or rot. They stack compactly. They take less skill to work with than wood. The steel itself is over 60 percent recycled. And being quickly assembled with wallboard screws, steel stud walls can be as quickly taken apart and reused. I've seen interior sheetrock-and-steel-stud walls for an office installed *over* the wall-to-wall carpet that was laid down first for the whole floor. Now that's casual.

It strikes me that external walls can take either a High Road or Low Road approach, encouraging either permanence or change. In some buildings you might want the front to be impressively High Road and the back adaptively Low Road. Low Road walls offer a further choice—they can be funky or high tech. One of the best of the funky is the highly forgiving board-and-batten. It doesn't need precise fitting and can ignore wood expansion and contraction from moisture and temperature. Anybody with a hammer and saw can make or unmake a decent board-and-batten wall. For high-tech Low Road walls, the current acme is the variations on Dryvit known in the trade as "exterior insulation and finish systems" (EIFS). Light and cheap, they are factory-made panels of stucco and foam insulation bonded together, shortcutting weeks of work in wall construction. They can be molded to decorative shapes and look right nice in a movie-set sort of way, at least for a while. If they show dents after a few years, or the bonding compounds loosen, it is no great loss to replace them, maybe with something more lasting.

High Road walls are nearly always masonry. Stone is grander. Brick is more adaptable. Since cavity walls for brick are now the standard, be sure the ties between the layers are stainless steel, or the permanent look may become a dangerous illusion in a few decades. Rammed earth has soundproof, energy-conserving two-foot-thick walls, but it's hard to change. An attractively cheap material related to masonry is stucco. Unlike the countless forms of fake stone cladding which always look corny, stucco is an ancient and honest material. If well crafted, it is impervious to weather and fire and needs little maintenance. It doesn't accommodate change as well as brick.

I favor keeping Services separate from Skin as well as from Structure. The cosmetic practice of hiding wires and pipes in walls makes maintenance and improvement a major hassle. The conservative tactic—at higher initial cost—of installing over-capacity electrical feeders and breakers, oversize chases, and an apparent excess of outlets is nearly always rewarded. The general rule is: oversize your components. Another adaptive trick—both for incremental and radical change—is to keep wiring accessible in what's called wire mold on the walls or in cable troughs hanging from ceilings. Some nice-looking baseboards and chair rails are made of hollow PVC to hide wiring accessibly. On our tugboat we elected to keep all pipes and wires easily reached and lost nothing aesthetically thereby. The tan-enameled metal wire mold is inconspicuous on the varnished wood walls, and exposed pipes look romantic on a boat.

Anticipate greater connectivity always. When Berkeley's Environmental Design Building, Wurster Hall, was built in the early 1960s, the whole building was laced with extra conduit in anticipation of coaxial cable for video classrooms. It never happened, of course. What happened instead was an utterly unpredicted infestation of personal computers, which eventually needed to be netted together. How convenient it was to find those empty TV conduits. All new buildings should have extra conduit laid throughout—two or three or more vacant half-inch plastic conduits with labeled string hanging out of the ends, waiting for unplanned phone lines, speaker leads, computer wires, coaxial cable, or what have you.

Anyone who buries services in the walls should adopt the practice of design/builder John Abrams in Martha's Vineyard, Massachusetts (who will emerge as the hero of this chapter). In all his buildings he has a ritual just before the walls close, when the services are installed but the sheetrock hasn't been put on yet. He walks methodically through the building photographing every open wall and ceiling and keying each photo to a set of plans. He discovered that the procedure gives him a chance for close overall supervision of the job at a critical moment, but the main purpose is to record precisely where all the services are before they get hidden—every duct, pipe joint, and outlet knockout. The color photos are then assembled with the keyed plans in a ring binder, which is known as "The Book" throughout the rest of the job. The

December 1935. Library of Congress. Neg. no. LC-USF 342-1159-A.

BACKYARDS ARE WHERE THE ACTION IS. In this 1935 photograph of Johnstown, Pennsylvania, nearly every house shows signs of additions and changes in the back. The fronts maintain a more High Road decorum. (This is a Farm Security Administration photo by the famed Walker Evans.)

1991. John Abrams

"27." This series of four photos (originals in color) shows the long interior wall of the living room. In this photo, electricians can see that there is enough extra wire in the wall so that the outlets might be moved several inches if necessary. And someone looking for a stud in the middle of the picture will know that it's not there because of the recess for the bathroom mirror cabinet on the other side.

"28." There's an intriguing cavity behind the wall above the fireplace that doesn't show up in the plans and is invisible once the wall is closed. I'd be tempted to put a sculpture niche or a wall safe or something in that space.

"29."

"30." A tall later owner of the house might be interested to know that there's enough solid wood over the arched doorway that it could be raised a few inches fairly simply.

PHOTOGRAPHING WALLS BEFORE THEY CLOSE has become a standard practice for design/builder John Abrams, because it makes later adjustment of the building so much easier. The photos reveal exactly where the Services go and what are the hidden Structural elements.

At my request, Abrams went back to reshoot the photos from the same angles after the Weinstocks had moved in. It's a little shocking to realize the degree to which Stuff takes over once a building is lived in, and Services and Structure disappear.

In the bay window on the other side of the doorway is a change evidently made during finishing on the house. What is marked on the plan as a "built-in desk" for the study has become a pleasant window seat.

1991. John Abrams

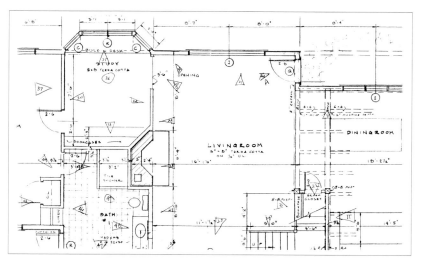

On the plan of the house, the photos are indicated by numbered arrows. Here, numbers 27 through 30 up the left side of the living room are the photos on the facing page.

The soon-venerated "Book" is made up of all the open-wall photos plus the keyed plans. The times of greatest value of the book are during finishing, later during remodeling, and any time the house is sold. In 1991 Abrams's company had made twenty such books since they began the practice in 1986.

Abrams keys the photos to the house plan (above) and assembles them into a book (right). The book is heavily used by subcontractors throughout the rest of the job and then is turned over to the new owners. The owners of this Martha's Vineyard home are Davis and Betsy Weinstock, both writers.

1991 - One side of the Weinstock house, showing the study's bay window and the many-paned window of the living room. I would bet that exterior photos like this, made soon after occupancy, could usefully be added to the book of open-wall photos. They provide a base for studying later effects of aging and deterioration, and subsequent owners might want to know what the original building looked like, before the inevitable additions and Skin changes.

electricians, finish carpenters, and other subcontractors consult The Book as often as they do the building plans. The electricians look to see if there's enough slack in the wiring to move a light switch; the carpenters want to know where the blocking is for nailing the trim. At the end of the job, Abrams ceremoniously turns over The Book to the owners. With the passing years it becomes ever more precious for repair and remodeling, as original memories fade and construction practices change. Eventually it becomes a major item in the resale value of the building. It costs almost nothing to make.

Abrams recommends doing the same thing for earlier stages, photographing the foundations and their drainage before fill-in, the septic system before fill-in, and all trenches. Everything that is buried will be dug up one day. It's better to leave a treasure map.

Can a building learn while it is being built? Here we come into the realm of theorist/architect/contractor Chris Alexander. He insists that architects can't really visualize how a building will look and feel, nor can anyone else—no matter how computer-enhanced they are—and so construction should be a prolonged process of cut-and-try. "Everybody wants to zoom," he says, "and you mustn't. You are constantly finding out about the building while constructing it, and what you will find out is inherently and necessarily unpredictable. You are watching a developing wholeness."

If that sounds extreme, bear in mind that Alexander is trying to shape buildings as superb as the works of living nature, and by the same method—ruthless evolutionary selection. For contrast, study ambitious buildings constructed in the normal way, where often blunder is layered upon blunder.

The third expansion building of the Library of Congress in Washington, the Madison Building (1980), was designed with a spacious atrium in the middle open to the sky. But late in the design process the Fire Department nixed the atrium, worrying about a dangerous chimney effect, so floors were built over the space, leaving a three-story-high dark and echoing void below, with a disconsolate fountain in the middle. Offices that looked into this space were provided with huge mirrored-glass windows so that workers could enjoy the view without themselves being visible. Thanks to the darkness of the covered atrium, it worked out just the opposite. The office workers could see only the reflection of themselves slaving away, and tourists (the atrium opened onto the entrance lobby) could study them as if they were in museum dioramas of 20th-century white-collar toil. Eventually the mirrored glass was replaced with clear, and everybody could see everybody, and visitors were kept away from the dreary core of what was meant to be a magnificent building.

In 1992 in Bonn, the Bundestag (Congress) was moving into a brand-new building with an enormous plenary hall symbolically round and transparent—the surrounding observation galleries were completely fronted by bulletproof glass. An elaborate computerized sound system was installed to eliminate problems of feedback and volume adjustment. At the first meeting in the new space—the discussion topic was cost overruns on the new building—the sound system turned itself down to an inaudible whisper. The meeting adjourned for five hours while frantic technicians searched for the problem.

It turned out that sound was reflecting so perfectly off the surrounding glass that the only feedback-free level the sound system's computer could find was a faint murmur. Germany's grandest public meeting room was so constructed that it could not be used for public meetings. The Bundestag returned to its former building. Among the ideas to salvage the new plenary hall was a proposal to cover the observation glass with carpeting.

Try things first. Feel your way.

Chris Alexander believes in making models. Not just one fancy presentation model but whole lineages of rough models, with everyone who will use the building getting a chance to study and critique the best of them as they evolve. Gradually the models grow in scale until they are life-size and at the site—chalk lines on the ground or the floors, tentative portions of the building jury-

1992 - PREMATURELY COMPLICATED, this $140,000 new house attempts to look as if it has been added on to for generations. As a result, actual add-ons will be difficult, and the fussy complexity greatly increases the construction and maintenance costs of the original house.

rigged of cardboard or cheap plywood. Opulent buildings like the Library of Congress or the Bundestag building need proportionally elaborate models to test for potential folly. Computer modeling and simulation can be helpful, but it can also misdirect designers toward illusory "optimal" solutions. These are inevitably brittle—they work only under one set of conditions. Computer-aided design (CAD) has led to weaker building that can't handle unusual circumstances such as hurricanes. Before computers, engineers dealt with uncertainty by building in a large safety margin. They were right. Oversize your components.

The cut-and-try approach can lead to doing some parts of a building quick and dirty with the idea that they will be temporary tests. If they work well, one imagines, they will be improved later. If they don't work, it's no loss to replace them with something that works better. This can be a wholesomely Low Road invitation to later refinement, but beware: in the real world "temporary" is permanent most of the time. If the cheap trial worked, it will be left alone, no matter how funky it is. If it failed, it's embarrassing to fix. Life rushes on to more pressing or interesting problems.

Once "finishing" on a building begins, the guessing and trying multiplies. Contractors know that when a building is fully closed to the weather, the job is only half done. A maddening amount of debate and adjustment is coming. Even so simple a thing as paint color gets complicated. An anthropologist friend of mine studied paint-purchasing behavior for an advertising firm. It turns out that people can't guess from the sample chips in stores how paint will actually appear on a wall. "They buy a quantity, go home and paint half a wall, groan at how it looks, and then they bring it back to change the shade, usually by lightening it." Try modeling that on a computer; it's worse than sample chips.

Some say you should flee a building while it is being finished or remodeled. I recommend occupying it. Bad as the inconvenience and aggravation get, it's worthwhile for the fine-tuning that only presence at the worksite affords. My wife and I lived in our tugboat's pilot house while the kitchen below was being built, and we lived in the pilot house and kitchen during the months it took to gut the engine room and turn it into a cozy library. Living with the finishing oil on the stripped wood walls of the pilot house bedroom let us know it was a failure—too sticky and dark. The kitchen and library got a very satisfactory antiquing varnish as a result. When it came time to figure the height of the tilt-up Corian cutting board, Patty stood with vegetable knife in hand and the contractor measured the distance from the blade to the floor. We did the same with the kitchen sink and the refrigerator, positioning both much higher than the norm, so we reach into the refrigerator without calisthenics and wash the dishes without backache. Our home fits us like tailored clothing.

Living at the construction site is an honorable vernacular practice. Limited resources made people occupy immediately, then build for years. Inside some of the nicest old adobe homes at Seton Village, near Santa Fe, you will find fully intact old railroad cars. That's where the original family lived while the adobe construction labored on around them.

It would be nice if architects would design tiny starter homes for people, but they won't. The profit margin is too small. And developers won't do it for fear of encouraging too much grass-roots autonomy and change. Nearly all homes that grow from modest beginnings are, like Thomas Jefferson's, owner-built and owner-designed. Usually an architect is rented briefly just to sign the plans. I think there's an opportunity waiting for professionals who can figure out how to design and construct houses that grow incrementally over time.

October 1939. Library of Congress. Neg. no. LC-USF-34-20922-C.

1939 - During the Depression, houses were often built by stages, as money came available. A tent on a floor platform like this is much nicer than a tent on the ground. The Farm Security Administration photo (taken near Klamath Falls, Oregon) is by Dorothea Lange.

October 1939. Dorothea Lange. Library of Congress. Neg. no. LC-USF-34-211438-E.

1939 - Sometimes cellars were built with the intention of living in them until enough money had accumulated to complete the house. The wait might be years, so they were built to be comfortable. This one in Nyssa Heights, Nebraska, had electricity.

LIVE/BUILD is the intimate, traditional way. Dwelling at the construction site can save money, can add to site-sensitivity and use-sensitivity in the design, and can either spread the construction costs and hassle out over time or goad the dwellers to hurry up and get the damned thing finished.

ca. 1985 - This two-stage house in Bayfield, Colorado, is fully livable from the very start of construction. The ten-wide mobile home has its own kitchen, bathroom, bedroom, and living room, so the construction of the standard house around it can go as painstakingly or as seasonally as the owner wants.

ca. 1985 - The potter-jeweler living in this 8-by-45-foot house trailer in Crawford, Colorado, added a workshop (with recycled windows), a greenhouse (left), and bales of hay for insulation (right).

Allan D. Wallis, from his book *Wheel Estate* (New York: Oxford, 1991), p. 153.

Paul Heath. From *Wheel Estate* (New York: Oxford, 1991), p. 156. See Recommended Bibliography

1993 - HANDMADE HIGHRISES like this are the contemporary vernacular for new housing throughout the eastern Mediterranean. There are tens of thousands of buildings just like it—concrete frame, structural clay tile walls (often made by the occupants), three to seven stories high (no elevator), often a solar water heater on top. Small local firms frame up the building in concrete with hand-built forms raised story by story—frequently with rebar left sticking out of the top in case more floors are added later. The floors are finished and occupied in random sequence (as here, in Seljuk, Turkey) because the floors are sold as condominiums, and the families complete their space and move in when finances permit.

3 June 1993. Brand.

Structural clay tile and mortar is the universal medium for new construction, even for some kitchen furniture, as here. Pipes are let into the tile by knocking in one or more cells. The surfaces of the blocks are scored to hold stucco finish. Unlike brick, the tiles are wide and light enough to be laid up only one layer thick for exterior as well as interior walls.

4 June 1993. Brand.

Everywhere I went in Turkey (this happened to be at Sardis) were stacks of locally manufactured structural clay tiles. They are an ideal incremental medium in a high-inflation economy (70 percent a year in Turkey). When you get a little money, buy tiles or build a wall. The weather will do no harm to it while you wait for the next bit of cash.

2 June 1993. Brand.

Buildings that last are made of quality materials and with quality craftsmanship. It costs. Lumber that is close-grained, straight, and well dried is expensive. Quality builders are expensive too, but the investment pays well later in terms of durability and flexibility. Fine artisans treat code requirements as setting a minimum standard rather than a maximum. They know that step dimensions in stairs have to be consistent to within three-sixteenths of an inch or people will stumble and get hurt. (A major source of architectural malpractice suits is for sloppy or overly creative stairs.)[6] Good carpenters employ measuring distinctions as fine as the "RCH."[7] Contractor Matisse Enzer explains: "RCH is a technical term, a unit of measurement an order of magnitude smaller than the last unit mentioned in a stream of adjustment commands—'Up an inch. Down a half. Down a quarter. OK, up an RCH and nail it.'"

An American carpenter apprenticing in Japan reported on their version:

> I watched four senior carpenters standing at attention, silently accepting a stiff rebuke from the master. Their crime: someone had miscalculated a few millimeters on a hip rafter. The difference was hardly noticeable, even close up, but since the beam was designed to achieve its perfect form only after several years of sagging and shrinking, this small error would be magnified and possibly distort the whole. Fumed Nishioka, "They'll laugh at me. They'll say, 'That's not the way a hip rafter should look!' And I won't be around to defend myself."[8]

It was a temple they were working on, but every building has some crucial parts that have to be done carefully for it to age well.

---

6   The optimal dimensions for stair safety are said to be a 7-inch riser and 11-inch tread.

7   Some say it stands for "royal cunt hair," some say "red cunt hair." Whether the usage will survive, given the growing numbers of women carpenters, remains to be seen.

8   S. Azby Brown, *The Genius of Japanese Carpentry* (Tokyo, New York: Kodansha, 1989), p. 67.

A CONTINUAL HOUSE—its first twelve years. Cabinet-maker (later building contractor) Stephen Seitz built a huge shell, quickly moved his family into one room of it, and then proceeded to finish the interior and refine the exterior for years. Some people start with a small, very well "cooked" (finished) house and then add to it. Some, like Seitz, start large and raw, and cook the parts gradually, to taste.

October 1980 - Near Otego in upper New York state, Seitz's family had a forty-acre parcel. For their home they picked this site overlooking a creek. The frost goes two feet deep in the winter. A basement is a necessity for a stable house.

August 1981 - Working mostly alone, Seitz framed up the basic structure. He elected to use 1-inch rigid foam insulation with aluminum foil on the exterior, and was glad he did. The foreground corner of the building he sided with logs. The bulldozer was grading the backyard.

December 1981 - All that winter, for a half day every day, Seitz nailed cedar shingles to the huge roof. He later wished he'd bought a higher grade. This is the back of the house—south-facing, lots of sun. Like all buildings, it will keep making adjustments to handle the sun.

September 1989 - When Seitz's contracting business prospered, he added a steel-roofed, three-car garage with room upstairs to store salvaged building materials. The greenhouse was completely rebuilt (and a hot tub installed), the logs on the left were rechinked with cement, and cedar siding made it halfway up the right side of the house. In 1988 Mrs. Seitz came home from birthing their third child to a completed master bedroom upstairs.

January 1981 - The basic plan of the house was taken from purchased architectural plans, which the Seitz family adapted to their own needs and ideas. This series of photos was taken by Seitz's friend (and early co-worker), Kevin Kelly.

May 1981 - Seitz began the flooring and framing with a romantic notion—to use locally-milled rough-sawn hemlock boards, save money, and support the local economy. But the lumber was too inaccurately dimensioned and a pain to work with, so he switched to boards and plywood shipped from the west coast.

September 1982 - The Seitz family of four (kids aged one and two) moved into a single room behind the greenhouse, heated with a wood stove. Seitz had a workshop in the basement. The rigid foam insulation now served (pretty well!) as the skin of the house. For years it was known in the neighborhood as "the aluminum-foil house." Chimneys went up for a wood stove (left) and massive fireplace (right).

June 1985 - Finishing work went ahead quickly on the left end of the house, where Mrs. Seitz's aunt was due to move in. She arrived in 1984 and requested a screen porch next to the log wall. The greenhouse was enclosed but proved to be too hot, its frames got warped, and it was used to store firewood. About this time Seitz became a building contractor (note pickup truck).

November 1992 - Seitz gave up on the logs (full of snakes and carpenter ants) and replaced them with salvage redwood siding. Fieldstone siding went up on the left gable end. A garden shed and swing set were added (far right), plus an enclosed garden (right), and a nylon screen for the greenhouse. Maples and white birches were planted, for eventual summer shade. Five rooms were still being finished in 1993, including a library, the living room, and the main hall.

All photos, Kevin Kelly.

Good artisans also know or can figure out the details that will make a building work right for the inhabitants, so the toilet-paper dispenser can be easily reached and the shower doesn't spray the bathroom floor or burn the hand that adjusts the hot water. Well-made buildings are fractal—equally intelligent at every level of detail.

But the whole building doesn't have to be gourmet-cooked; some parts should be left raw. Chris Alexander routinely cites the authority of the Japanese artistic virtue of *wabi sabi*—"the recognition that in a beautiful thing there is always some part which is lovingly and carefully done, and some parts which are very roughly done, because the compensation between the two is necessary in a real thing." Undifferentiated, unpartitioned, unspecialized loft, attic, basement, garage, or storage spaces are essential.

"Finishing is never finished," but at some point you have to just stop, let the builders go away, and start living in the place. Inhabitation is a complex, thrilling, prolonged event. The way to treat a new building is as if you might move away tomorrow and as if you expect to be there the rest of your life. Inhabit early, build forever. Part of relishing a place is tinkering with it. You're going to have to anyway, because no building works right at first. It will take a year to work out just the major bugs. In the first year of the Media Lab building at MIT, the elevator caught fire, the revolving door broke weekly, all the doorknobs in the building failed and had to be replaced, the automatic door closers were stronger than people and had to be adjusted, and an untraceable stench of something horribly dead filled the public lecture hall for months. This is normal.

John Abrams, the design/builder in Massachusetts, discovered that his business was being hurt by new-building problems on the far side of the "occupancy" barrier. "We would have these terrific relationships with clients," he recalls, "and then all of a sudden they'd start to deteriorate right after the completion of a building, due to very minor problems. My thinking always was, we've been on this grand excursion with these people, and we've done such a great job, and we love them and they love us; these little things don't matter that much. Well, they turned out to matter a lot.

"After seeing this for a while and seeing that it was fundamental, we hired someone just to handle those problems—a full-time handyman for our clients. He does nothing but adjust, repair, and maintain, and he's always available in a day or two. We never question whether a problem is our fault or their fault or anybody's fault. It's something that's wrong with the building, and it has to be dealt with immediately! We don't bill for much of his time. We believe it's our duty in building a building that it should work well for twenty-five or fifty years. We're trying to make a building that doesn't need his time, and whatever time of his it needs is our responsibility. That completely changed how our relationships with our clients endure. When they eventually want to improve or change the building, they come back to us. That's where a lot of our work is now."[9]

The follow-through also has the effect of training both the builders and the clients, Abrams says. "We get to see what works in our buildings and what doesn't work. Our crews have become adapted to really dealing with all the minute details before they become a problem." Worried clients learn from the handyman about some new-building problems that will simply go away with time. "They'll have moisture problems," says Abrams, "because a new building is *loaded* with water. In the concrete, in the lumber, in the plaster, in the paint—hundreds of gallons remain in the building, and it takes a year or two before that's all really dried. The problem will disappear."

---

9 The brochure for Abrams's company, South Mountain, states, "It is important to us that the relationships we establish—with clients, subcontractors, suppliers, municipal officials—be regarded as integral parts of the building procedure. The design and construction process is long, sometimes arduous, and full of detail. Supported by respectful relationships, it becomes a source of pleasure." In further pursuit of continuity, Abrams restructured the business as an employee-owned corporation, so workers are motivated to stay on with the company and refine its skills. The three foremen have become active site-based participants in the design of the buildings.

When he turns over a new house to the family, Abrams writes a detailed letter that goes with The Book of photographs of the open walls and a packet of manuals of all the operating equipment in the house. The letter says, "During the first year, we will deal with all subcontractors. At the conclusion of the first year we can do a walk-through and try to solve any problems that are unsolved at that time." In the letter are the names and phone numbers of all the subcontractors on the building, along with a suggested schedule of maintenance.

At the building site Abrams has the work crew leave a tidy pile of lumber ends that can be used as firewood, plus surplus building materials such as bathroom tiles that will make later repair or expansion easier. Historic building maintenance professionals have a standard list of materials worth stockpiling—such things as extra shingles, tiles, or slates, roof gutter materials, bricks or building stone, paving materials, awnings, door hardware and doors, window hardware and windows, specialized glass, paneling and trim, and wallpaper.

Besides spare parts, a new building needs a complete and accurate record of itself. A building of any complexity must have a set of "as-built" drawings showing what was actually constructed. Rehab and restoration architects routinely charge significantly less for their work if as-builts are available. Part of the shame of the headquarters building (1970) of the American

Institute of Architects in Washington is that it was completed without as-builts, making the later all-too-necessary renovations even more of a pain. There also needs to be assembled in one place a complete legal paper trail of the building—all the contracts, bonds, guarantees, easements, covenants, zoning requirements, etc., plus title records and an address list of every official and lawyer dealt with. This collection will grow.

Maybe the right place to stash it would be in a little shrine—Roman style—to the genius of the domus. In a fireproof box, of course.

All ships have scrupulous maintenance logs, and so should buildings. The logs show precisely what was done, when, and by whom. They also schedule the routines of periodic inspection, servicing, and preventive maintenance. In buildings the record of work done can sometimes be attached to the point of work, such

WRONG DOOR. Think of every restaurant, shop, and public building you've visited. The entrance has double doors, by law. But one door opens and one doesn't, and you can't tell which is which until you've crunched into the wrong one. That one detail of staff failing to unlock both doors shrieks of laziness, disinterest, and unwelcome. Every customer enters in a state of having been humiliated by the building, by a nuance untended to.

16 December 1990. Brand.

1990 - "Please use other door." *Which* other door?! This is the Folk Art Museum in Santa Fe. I watched visitors try one door after another (the middle one is locked too) before entering in wrath by the left-hand door—against the deeply internalized convention of going to the right in two-way traffic. The doorway slogan says, "We are all one." The doors themselves say, "Go away."

1993 - California architect Julia Morgan was trained at the École des Beaux Arts in Paris. All of her buildings have the fractal quality of offering delight at every scale. This building in San Francisco, called The Heritage (1925), shows the same degree of complexity as the tree next to it. The brick detailing and terra cotta trim on the west-facing facade change with the sun all afternoon The bay windows mark the public rooms within. Of this home for the elderly, Morgan's biographer writes, "The enthusiasm of the residents and long waiting list testifies to a carefully conceived and executed environment for the aged that has rarely been equalled."[11]

FRACTAL BUILDING. The inventor of fractal geometry, Benoit Mandelbrot, has an explanation for people's dislike of grossly pure shapes like the Seagram Building in New York. According to James Gleick, "Simple shapes are inhuman. They fail to resonate with the way nature organizes itself or with the way human perception sees the world.... Against the Seagram Building [Mandelbrot] offers the architecture of the Beaux-Arts, with its sculptures and gargoyles, its quoins and jamb stones, its cartouches decorated with scrollwork, its cornices topped with cheneaux and lined with dentils.... An observer seeing the building from any distance finds some detail that draws the eye. The composition changes as one approaches and new elements of the structure come into play."[10]

People are happiest also in buildings where change occurs at every scale from weeks to centuries. Such buildings are fractal in time.

as servicing tags on equipment. A room's paint, wallpaper, and carpet specifications can be taped to the inside of a light switch cover in the room. Old lead roofs customarily have the date of their installation marked in them, along with the name of the "plumber" who did the work, and contemporary lead roof renovators continue the tradition.

In a healthy building, maintenance, correction of faults, and improvements all blend together. Chris Alexander has the best-developed theory of how it works. He says that a good building undergoes progressive change until it reaches an adapted state, and this is accomplished through a progress of whole states driven and shaped by individual projects:

> An organic process of growth and repair must create a gradual sequence of changes, and these changes must be distributed evenly across every level of scale. [In developing a college campus] there must be as much attention to the

repair of details—rooms, wings of buildings, windows, paths—as to the creation of brand new buildings. Only then can an environment stay balanced both as a whole, and in its parts, at every moment of its history.[12]

It's a waste to do anything for just one reason. Besides serving its immediate purpose, Alexander suggests, each project should serve a larger goal of "healing the whole."[13] And it should prepare the way for a larger and more significant whole. The addition of a porch—enjoyable in its own right—might correct a problem of excessive sunlight on the southwest side of a house and at the same time invite (but not require) the later planting of a garden around a path leading from the porch steps. Alexander adds that each project should also "generate smaller wholes in the physical fabric." Let the porch steps be a pleasant place to sit. See if the area under the porch might work as a place to store yard tools. Could the railing be wide and strong enough to perch on?

Equally, the removal of a decayed porch which is unused because of a chilly climate might suggest the addition of mudroom for the heavily trafficked back door, plus a bench in the sun on the side of the mudroom that is protected from the wind. Learning in a building, then, is a simultaneous process of constant self-healing and of arranging for greater possibilities.

Nuances are as important as systemic problems. A roof leak or a constricted hot-water pipe will cause increasing damage and must be fixed, but so must the door hinged on the wrong side or the window you want to open but can't. Fine-tuning is what turns a building from a nuisance into a joy.

The point is to make adjustments to a building in a way that is always future-responsible—open to the emerging whole, hastening a richly mature intricacy. The process embraces error; it is eager to find things that don't work and to try things that might not work. By failing small, early, and often, it can succeed long and large. And it turns occupants into active learners and shapers rather than passive victims.

Loved buildings are the ones that work well, that suit the people in them, and that show their age and history. All it takes is keeping most everything that works, most everything that is enjoyed, much of what doesn't get in the way, and helping the rest evolve. That goes better if the place is neither owned nor maintained by remote antagonists, because they distance the building from its users. What makes a building learn is its physical connection to the people within.

Finally, an adapted state is not an end state. A successful building has to be periodically challenged and refreshed, or it will turn into a beautiful corpse. The scaffolding was never taken completely down around Europe's medieval cathedrals because that would imply that they were finished and perfect, and that would be an insult to God.

What about the building you see when you look up from this book? Go do something timely to it.

———

[10] James Gleick, *Chaos* (New York: Viking, 1987), p. 117.

[11] Sara Holmes Boutelle, *Julia Morgan Architect* (New York, Abbeville, 1988), p. 120.

[12] Christopher Alexander, et al., *The Oregon Experiment* (New York: Oxford, 1975), p. 68. The book reported on the inventing of a dynamic master plan for the University of Oregon in Eugene. See Recommended Bibliography.

[13] Christopher Alexander, et al., *A New Theory of Urban Design* (New York: Oxford, 1987). See Recommended Bibliography.

APPENDIX: The Study of Buildings in Time

ALL OF THE BIOLOGICAL SCIENCES make sense—and make sense of each other—in the light of one unifying concept, Darwin's theory of evolution. Something similar could unify the disciplines, professions, and trades that have to do with buildings. They could become, like biology, one organic body of knowledge and inquiry. The missing link is time.

Architectural historian Patricia Waddy prefaced her study of 17th-Century Italian palazzos with the observation,

> Buildings have lives in time, and those lives are intimately connected with the lives of the people who use them. Buildings come into being at particular moments and in particular circumstances. They change and perhaps grow as the lives of their users change. Eventually—when, for whatever reason, people no longer find them useful—they die. The artistry of the designers of buildings is exercised in the context of that life, as well as the context of a life that art itself may have.[1]

The statement suggests a deeper role for the building professions and disciplines. Architecture has trapped itself by insisting it is "the art of building." It might be reborn if it redefined its job as "the design-science of the life of buildings." A shift that minor could transform the way civilization manages its built environment—toward long-term responsibility and constant adaptivity.

Architects are almost ready, almost desperate enough, to make the shift. In 1993 the new president of the American Institute of Architects, Susan Maxman, declared, "The profession pays far too much attention to little gems of new buildings. If we are to survive, let alone prosper, we must fundamentally retool. We must equip ourselves with new kinds of knowledge, skills, and attitudes which will support our work as renovators."[2] She called on her profession to move toward sustainable design. In June of 1993 the largest-ever gathering of architects convened in Chicago for the eighteenth World Congress of the International Union of Architects. Its theme was "Architecture at the Crossroads: Designing for a Sustainable Future." *Sustainable* is a buzzword, meaning "ecologically correct," but it does stimulate thinking toward durability and open possibilities.

Among academic architects two other potentially useful time-honoring terms have been in occasional use—*synchronic* and *diachronic*. Linguists invented the words to describe two ways of studying the history of a language—the way it all fit together at one point in time (synchronic), or the way it developed over time (diachronic). Architectural historians adopted the same usage. Buildings of the past can be studied in terms of how they worked and interacted at one time (the preference of city planners and architects looking for design ideas), or in terms of how they evolved over time (the preference of architectural historians). A common criticism of the historians is that they should study the past the way designers study the present—synchronically, in terms of immediacy. I would make the opposite argument, that designers should study the present the way historians study the past—diachronically, in terms of change over time.

---

1 Patricia Waddy, *Seventeenth Century Roman Palaces* (Cambridge: MIT, 1990), p. xi.

2 Quoted in Nancy Levinson, "Renovation Scorecard," *Architectural Record* (Jan. 1993), p. 73.

3 Bernard J. Frieden and Lynne B. Sagalyn, *Downtown, Inc.,* (Cambridge: MIT, 1989), p. 315. See Recommended Bibliography.

Diachronic understanding and design is already well established and at work in city planning. That success should inspire and educate the rest of architecture, says Frank Duffy: "The biggest artifact we put together, the city, is actually quite good at accommodating change. Perhaps because of its size and complexity, we have learned not to own it personally as designers and to tolerate its changes. Cities mature. Cities flex through time." The recent crop of neotraditional town planners are comfortable with adaptive features like the revived alley and the roll curb. Alleys through the middle of residential blocks, they rediscovered, work like basements—they separate Services from Structure, and they separate formal front-of-the-house activities from faster-changing informal activities in back. Gentle roll curbs on the streets make a new curb cut unnecessary if you want to move or change a driveway.

The whole profession of city planning has been more responsive than architects to pressure from users. Following the outrages of "urban renewal" in the 1950s, neighborhood groups organized, took power politically, and hired the planners to work for them. The result is summarized in *Downtown, Inc.*:

> Development strategies have come a long way since then: from bulldozing whole neighborhoods to practicing microsurgery; from compulsive modernization to preserving a sense of the past; from designing office and apartment complexes for isolation to creating attractions that draw the crowds; from pushing city solutions on developers to solving problems through negotiation; from raiding federal highway and renewal budgets to packaging local and private sector funds.[3]

City planning used to imitate architecture, and it failed because of that. If architecture now began to imitate city planning, it could learn to succeed better. The many architects who left their profession to become city planners need to come back, bring what they learned with them, and start designing buildings that

flex and mature the way cities do.

The comparison of what happened in city planning versus what happened in architecture could be the beginning of a productive self-critique by the architecture profession. Students would love it. What made Architecture allergic to time? What made Architecture afraid of building users? How did style obsession and the star-architect system manage to keep redominating the profession? Get down to cases—what exactly is the performance record of buildings that won architectural awards? What is the effect of rewarding bad performance? Could vernacular design be rethought? How did architects get away with "evoking," "referencing," and "bowing to" local vernacular traditions instead of learning from them? What would it take for magazines like *Interiors* and *Architectural Digest* to stop drooling over luxury interior design and start investigating what makes rooms really work?

To completely reunderstand buildings would require both of the fundamental approaches to knowledge—observation and theory. I called them "Look first" and "Think first" in a seminar I led on "How Buildings Learn" for architecture students at Berkeley in 1988. On the last evening I handed out blue books and explained that there were two types of people in the world—those who deal with something new by really looking at it, devoid of preconception, versus those who prefer to form hypotheses first and then study the thing to see which ideas were right. Both are honorable and productive. With "Look first," new perception changes understanding. With "Think first," new understanding changes perception.

I asked the students to identify their preference and join with others of their kind on one side or the other of the room. The exercise for the two groups was to write in their blue books the insights that might be found in a series of photos I unveiled—a sequence of color snapshots I had made of the class at the beginning of each of our nine sessions. The Look-firsters had to study the photographs immediately, while the Think-firsters

19 January 1988 - CLASS SNAPSHOTS. For no good reason other than random curiosity about photographs in sequence, at every meeting of my seminar on "How Buildings Learn" I took pictures of the students at the begining of the class. This was the first meeting.

1 March 1988 - At the last meeting of the seminar (partially shown here), I asked the students to analyze the sequence of photos of the group. One student, who took the theoretical rather than observational approach, wondered why the women in the class wound up clumped by the door on the right.

stayed in their seats writing down what they thought they might find in the pictures. Then the Look-firsters would sit down to write their reports while the Think-firsters finally got a look at the data. After both groups had written up their insights, they swapped blue books and wrote comments on each other's reports.

Much giggling broke out at that point, most of it from Thinkers reading the reports of Lookers, because Looker observations were so original they were comical. "Every chair has its desk on the right side, and the pattern of people in the photographs is like a grove of trees in a strong wind—everyone is leaning toward their right." "There are the least crossed legs in the first and last class." But the Thinkers had more to write, and there was more follow-through in their method. One, who had hypothesized that "Over time, people will sit closer together and more to the front," wrote gleefully, "Wrong! People move away from the front over time." She went on to theorize, "It may be that proximity is no longer felt necessary to support the group process. People grow to trust the interaction." She inquired further, "Why did the women end up near the door and the men on the other side of the room? Clumping by sex?"

How Buildings Learn is an example of the Think-first approach (lacking exactly the originality and new directions that Looking-first could generate). It speaks mainly to theory—to moving architecture toward operational strategy, away from stylistic

interpretation, and toward immersion in the previously ignored effects and use of time. The shift from studying what buildings *are* toward what they *do* is fundamental, but the industry can gain greatly by it, at modest cost. As corporate strategists point out, "A world-class researcher costs one-half a lawyer."

Still, theories mislead as much as they lead. An MIT teacher in animal physiology drilled into his students, "The animal is always right. When in doubt, ask the animal." Look-first analysts of the effects of time on architecture have an excellent tool for "asking the animal" in the prodigious amount of photo documentation of buildings that exists, dating back to the 1860s. Even the most romantic researcher of buildings, John Ruskin, was impressed: "Among all the mechanical poison that this terrible 19th century has poured upon men, it has given us at any rate *one* antidote, the Daguerreotype."[4] Sequential rephotography of buildings already fills a considerable collection of little-noticed books.[5] But they were created haphazardly as hobbies; the practice could be far more systematic and revealing.

The small amount of rephotography I did for this book demonstrated how much fun it is. To step into the exact point of view of an old photograph is to step into a time machine. While I was happily perusing thousands of old pictures in various archives, I gradually learned what makes photographs most usable for sequential study. A building exterior photo shows the most clarity and unshadowed detail if shot in hazy or overcast weather, preferably in a season when foliage doesn't hide half the building. Vehicles or people in the frame instantly announce the era of the photograph, because everyone can intuitively date styles in clothing and vehicles. It is helpful when photos show more than just the building itself, since often it is change or lack of

---

4 Wolfgang Kemp, *The Desire of My Eyes* (New York: Farrar, Straus, Giroux, 1990), p. 158.

5 Some of the best are evaluated in the Recommended Bibliography.

6 Francis Duffy, "Measuring Building Performance," *Facilities* (May 1990).

August 1888 - SUMMER VEGETATION hides much of the detail of the Fairbanks House in Dedham, Massachusetts (see the series of photos on p. 121). Plantings around buildings are often photographed in fullest bloom because the dwellers or owners want to show off how nice they look.

Charles B. Webster. Society for the Preservation of New England Antiquities. Neg. no. 15797-B.

Winter 1888 - The same building in bleak winter reveals a great deal more—window detail, the poor state of repair, sagging of the shed addition at center-left, even the extent to which the massive central chimney appears to be holding up the whole house.

Society for the Preservation of New England Antiquities. Neg. no. H-14761.

change in relation to the setting that is most interesting. Since detail is everything, the larger-format the camera, the better. To my shame I did all my rephotography with a 35mm camera, though at least it had a shift lens to make vertical lines parallel and keep buildings from looking distorted.

Interior photographs are all too rare, which is a nuisance since that is where the most change occurs. Particularly rare are photos of the really hard working, rapidly changing rooms—kitchens and bathrooms. Rarest of all are pictures of the undefined spaces—basements, attics, garages, and storerooms. Interior photographs taken with available light best give the feeling of the place, but flash photos have a lot more detail. When I could, I shot both. People in interior shots can be distracting (because they are so interesting), but if the picture is taken informally, they help show how the space is being used. Interiors change so rapidly that to be useful for analysis, sequences need to be shot every few months or weeks, or even hours. For that sort of project I found it helpful to put a mark on the floor so that later photos could be taken from exactly the same angle, which aids comparison.

All photos of buildings should have the date of exposure penciled on the back of the print. For that matter, all photos of anything or anybody should have a date on them. Photographs are evidence, memories, history—peerless records of how something actually was. Precise dates multiply their value.

There is a shocking lack of data about how buildings actually behave. We simply don't have the numbers. To ever get beyond the anecdotal level—typified by this book—will take serious statistical analysis over a significant depth of time and an adventurous range of building types. Frank Duffy berated a conference of facilities managers, "What we do in the way of systematic measurement of building performance is a tiny drop in a sea of ignorance and indifference. Sloppiness is everywhere. Our results are too private. We measure what is easy to measure and ignore what is difficult. Real issues such as the use of space through time, productivity, and environmental responsibility thus tend to get ignored. Given the chance, facilities managers will retreat into the tiny box from whence they came—into neatness, housekeeping and a quiet life."[6]

There's no reason to rely only on professionals. This is not astrophysics; everybody is an expert on buildings. Amateur birdwatchers in their legions have profoundly influenced ornithology, population biology, and environmental politics.

TIME ANALYSIS of buildings on the scale of decades and centuries is well established in Europe, as in this survey of a 17th-century Gloucestershire farm building. Similar studies could be made of contemporary buildings on a time scale of days, months, and years. Patterns and paces in moving furniture, for example, might suggest what people really would like parts of buildings to do.

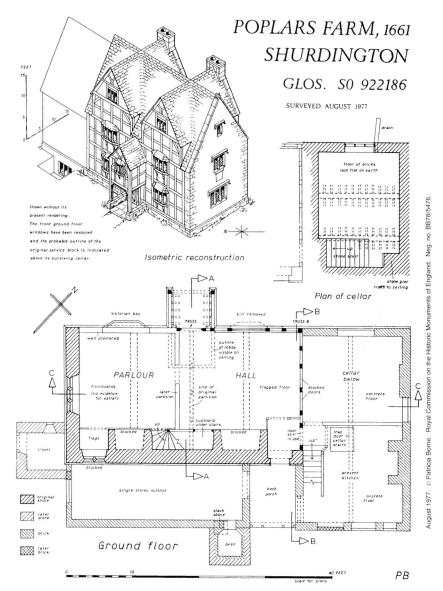

POPLARS FARM, *1661*
SHURDINGTON
GLOS. *SO 922186*

SURVEYED AUGUST 1977

Shown without its present rendering. The front ground floor windows have been restored and the probable outline of the original service block is indicated above its surviving cellar.

Isometric reconstruction

Plan of cellar

Ground floor

August 1977. © Patricia Borne. Royal Commission on the Historic Monuments of England. Neg. no. BB78/5476.

1977 - Buildings sleuths, some of them amateurs and hobbyists, love to reconstruct the sequence of changes in venerable buildings. What might they accomplish if set loose on contemporary change in contemporary buildings? This drawing (one of four pages) is by Patricia Borne.

Maybe amateur building watchers could do the same for the yet-unnamed science of building behavior.

The need is to study all kinds of buildings and all kinds of uses, not just the prestige ones or the high-revenue ones. The point is both objective—knowledge for its own sake in an area of shocking ignorance—and subjective. Our buildings are such disasters that we need detailed evaluation and rethinking at every scale. We need failure analysis that is systemic over the full scope of building-related activities and the entire life of buildings. When we investigate a building that is loved or loathed, it is not a question of praising or blaming the designer or owner but of teasing apart the whole tangle of relationships that make the building work or not. All buildings have problems, but only some correct the problems. What is the systemic breakdown that leaves a problem unnoticed? Or if noticed, unreported? Or if reported, unacted on? Or if acted on, unsolved?

Without that kind of corrective feedback a building can't thrive. Neither can the building professions and trades. A skilled carpenter told me, "I do things better now that I'm older. Why? Because I've spent so much time fixing the things that I did when I was younger."

The research needs academic rigor, but not academic irrelevance. A utilitarian bias should color the collection of information, the testing of ideas that emerge, and the transmitting of the ideas that work. For that reason it's worth taking a purposeful look at how knowledge transmission has worked in the past—how *do* buildings learn from each other? Cross-cultural study could lend perspective and also be a fount of ideas. I'm acutely aware that a book like this one would be radically different if it expressed

European or Asian experience—and what a joy that would be to research.

By considering buildings whole, university architecture departments could reverse their trend toward senescence. They could invigorate the faculty with an infusion of facilities managers, preservationists, interior designers, developers, project managers, engineers, contractors, construction lawyers, and insurance mongers. The departments could promote some of the marginalized people they already have—building economists, vernacular building historians, and post-occupancy evaluators. In that enriched context, what's left of art-oriented architecture would have all that its creativity could handle exploring new syntheses of the flood of data and ideas.

For such research to prosper at either the academic or the grassroots level, data has to cohere to buildings. We need to switch from the hotel-room aesthetic to the mountain-hut aesthetic about the accumulation of information. A hotel room is constantly scoured of any trace of previous use and is presented daily as if brand-new. Mountain huts are exuberant museums of their own past, with each hiker adding comments to the guest book, initials and dates to the woodwork, and food to the larder. What if every commercial building had an on-site journal and maintenance log, which the landlord could not legally remove or amend? What if city halls provided a repository for the full records of every house in town—not just the legal and title records, but photos and memorabilia voluntarily left by successive generations of tenants.[7] (I know from my research for this book that photo archives in libraries are incomparably more usable when organized by building or street address rather than by date of acquisition or name of collection.)

There are so many questions worth exploring. What are the oldest buildings in various cities that still command high rents? What made that happen? What kinds of buildings were torn down, and why? What is the distribution of building types in a city, and how does their longevity sort out? How about in small towns? What is the real distribution of design approaches—how many buildings are specially architected, versus franchise cookie-cutter, versus developer assembly-line, versus vernacular?

A prime opportunity for comparative study would be the uniform arrays of buildings all constructed at once, such as in the original Levittowns. What was the pace and range and process of their subsequent divergence? Architects like to think that upscale suburban developments like Greenbelt and Hollin Hills near Washington haven't changed much since they were built, and that's supposed to be a measure of their success. Is that true, or is the rapid change in the humble Levittowns an indication of the owners taking charge of the buildings and gaining greater satisfaction than the more passive tenants of Greenbelt? Which was the better investment?

What would an architecture student learn from accompanying a remodeling contractor through a series of jobs? How about following a building inspector around for a few weeks? Is there pattern to the realities they encounter? How does it match with the conventional wisdom taught in school?

What might be learned from highly detailed longitudinal studies of buildings in use? What changes from hour to hour, day to day, week to week, month to month, year to year, and over decades? This kind of study is the norm in ecology and some of the social sciences; there's no lack of lore about how to do it.

It would be interesting to investigate what happens with sequences of buildings that might be learning from experience. Dozens of Hyatt Regency hotels have been built around their trademark cavernous atriums. Are the later ones improvements on the originals? Did the originals later get adjusted to take advantage of what proved effective in the later ones? How is that

---

7   Boston architect Bill Rawn commented on this point: "The Boston Public Library became the depository for all buildings plans submitted to the city. It is an extraordinary resource for anyone planning to change any building. It confirms everything you've said in the record-keeping realm."

December 1988 - After two years in business, Phelan's was still in its original one big room, with mail order shipping on the far left, mailing on the far right, and customer service (taking phone orders) on the near right. This series of photos is the reverse view of what's on pp. 30-31.

21 March 1991 - Two-plus years later, the customer service area was wholly different. It had all-new furniture and filled the foreground. Since the adjoining bay behind the door had been taken over, the former hall into it had become a dressing room for customers, a bathroom door was added on its right, and a door on the left had been built and then closed up.

MONTHLY PHOTO-STUDY of a high-volatility work environment revealed patterns of furniture migration and shifts of workgroup boundaries. Constant change was both necessary and easy for Phelan's—an equestrian mail order catalog in Sausalito, California. Necessary because it was growing from $0.5 million/year revenues to $3 million in two years. Easy because it had cheap recycled furniture in cheap, roomy Low-Road space (a leftover World War II shipyard building).

25 September 1991 - One month later, the display shelves were back. And customer service desks on the right were beginning to encroach back out into thoroughfare space— each person on the phone was trying to get away from others so they could hear better.

13 November 1991 - Two months later, the printer has been replaced by a dog (food and water dishes, plus rug)—two burglaries had encouraged the adoption of a German shepherd for security. The new desk and dividers on the right came almost free from a software company that had moved.

21 June 1991 - Three months later (I was taking monthly photos from 22 locations in Phelan's, always at 2:30 pm), the desks in the right foreground had shifted around, and the printer on the far left acquired a cover. The display shelves formerly next to it had been replaced by a clothes rack for a summer sale. (The showroom of products for walk-in customers had grown from the distant right to include the area around the dressing room.)

23 August 1991 - Two months later, the clothes rack had gone, and the desks for customer service had retreated to the right. Partly they were pushed back by the intense traffic in the area—customers coming from the right, plus the growing number of staff walking from new space to the left and from upstairs behind the camera to use the company's only bathroom (center, dark). Also some customer service people had moved into new space elsewhere in the building.

Judging by these photos, office workers like to move their furniture much more than they're allowed to in most work environments where the space plan and management are too restrictive, or the furniture is too heavy. Constant, searching micro-adjustment is both empowering and adaptive. The boundary of a workgroup will flex back and forth between local and organization-wide needs.

21 March 1992 - Four months later, the pace of change had slowed. The nation was in a recession, and Phelan's had stopped growth for better profitability. The customer service desks continued to encroach outward.

22 October 1992 - Seven months later, the dog had acquired a cushy mattress and her own dividers. Approaching the peak season of Christmas orders, management required that there be a centralized workspace (center right) for all customer service "problems"—back orders, call-backs, returns, etc. Display shelves and a tall divider protected the privacy of the work surface—the very same piece of furniture that occupied that space four years before in 1988.

body of experience expressed to the designers?

Conversely, what keeps proven good new ideas from being widely utilized? What are the mechanics of blind conservatism? (And what are the advantages of blind conservatism?)

Another non-event worth scrutiny is buildings that don't change. Is it because they're perfect as is, or revered as monuments, or administered from too great a distance or through too many layers of bureaucracy, or is it that the occupants are old and set in their ways, or they can't afford change, or the material is physically impossible to alter, or what?

What are the usages that nourish buildings? And which ones destroy buildings? And which preserve them intact for revival later?

Suppose the kind of close study expended on shopping malls were applied to a couple of barrios. What does their rampant improvisation have to teach formal design? What does their responsible illegality suggest about amending property laws?[8]

Ships are the best-documented large structures in existence. Might they be studied simply as buildings? Could some of their high-density design and rigorous servicing discipline be transferred to ordinary buildings?

This book and others are full of unsupported hypotheses that would not be hard to prove or disprove. Are Modernist buildings that are designed "inside-out" really less adaptive than more traditional buildings designed "outside-in"? Find a few classics of each type and compare their histories. Does greater adaptivity go along with greater maintenance, or the opposite? Watch some low-maintenance and some high-maintenance buildings and see what happens.

How important is local control? An intriguing project would be to design some houses and small commercial buildings specifically to be serviced and maintained by their users. Only amateur skills would be required, and everything that needed work would be self-obvious. No outside expertise or special materials necessary.

Maybe this is the secret of high longevity at low cost—like the Volkswagen beetle.

Architects talk about "daylighting as formgiver" and "sunlighting as formgiver." What kind of buildings might reflect "time as formgiver"? If my adaptation of Frank Duffy's layering of buildings into Site-Structure-Skin-Services-Space-plan-and-Stuff is correct, how might building design better acknowledge and take advantage of that? All buildings grow, but some grow more outward and some more inward. What are the different advantages of the two paths, and how might initial design serve them best? Since vernacular buildings seem to show exceptional hardihood over time, could there be a subdiscipline of *applied* vernacular studies that would inform the cutting edge of design theory?

Then there's social context. What happened in Britain and America that made building preservation suddenly an irresistible force? What might make that transition happen in the city-trashing go-go economies of Southeast Asia? (Singapore, which adopted preservation almost overnight, would be an illuminating study.) How is the nature of human organization changing, and how are buildings reflecting that? A history of the office in this century would reveal a great deal. So would a history of the family.[9]

---

[8] In Peru's barrios, writes Hernando de Soto approvingly, "First, the informals occupy the land, then they build on it, next they install infrastructure, and only at the end do they acquire ownership. This is exactly the reverse of what happens in the formal world, which is why such settlements evolve differently from traditional urban areas and give the impression of being permanently under construction." Hernando de Soto, *The Other Path* (New York: Harper & Row, 1989), p. 17.

[9] A jolting view of the present is in Sherry Ahrentzen, *New Households, New Housing* (New York: Van Nostrand Reinhold, 1989). "The fastest growing household type is the single person living alone; persons living alone comprise 24 percent of all households. Single-parent families account for 12 percent. America's 86.8 million households are still dominated by the 50.3 million families maintained by married couples. Yet even within the conjugal family, lifestyle changes have occurred. Over 60 percent of married women with dependent children are in the paid labor force, compared to 18 percent in 1950. Nearly 53 percent of married women with children under 6 years of age are employed. Only 10 percent of households consist of an employed father, a homemaker mother, and children younger than 18."

All of these are pieces of a larger question: how does procedure learn? How do organizations adapt; how does design improve; how do strategies become robust? What differentiates that kind of self-improvement from mere succession? Some theorists say that any entity that learns must have a model of itself that can store past lessons and make future conjectures, but others say that's not necessary or may even cripple the immediacy of real learning. Who's right? Surely there's a whole taxonomy of subspecies of learning worth differentiating—unconscious learning, accelerated learning, wrong learning (superstition), goal-directed learning, aversive learning, situated learning, pseudolearning, creative forgetting. Organizational learning has been dissected by theorists Gregory Bateson and Chris Argyris into three levels: adjustment to a norm, change of the norm, and change of change. (This is seen in the pedagogic sequence of learning a language, learning languages, and learning to learn languages and hence anything.) Might that layering be reflected in a building designed for multi-level adaptivity? Maybe *that's* what Architecture Department buildings should be designed to handle—triple-level learning.

Since every building is expected to reach out thirty to one hundred years into the future, it's astonishing that the building industry doesn't do extensive futures research. Perhaps it's a paradoxical effect of the acceleration of change in our lifetimes. The very compression of events that makes futures study more necessary makes us more ahistorical in our outlook. We're too immersed in the onrush of change (no one calls it progress these days, interestingly) to do much looking backward or forward, and so we put ourselves at the mercy of events, repeatedly surprised and baffled. Buildings might help us out of that impasse. Their material persistence steadies us. Embodying the past, they invite us to think seriously and confidently about the future.

What, for example, might architectural futures-study focus on in the mid-1990s?

Some of the future is already in the pipeline—inevitable unless there's an asteroid collision. The most certain and influential of these is age distribution in the population. While the developing world (formerly called Third World) is getting radically younger, the developed world is getting markedly older. Older people prefer familiar buildings, especially in a time of disconcerting change, and they can overwhelm youth's desire for something new by the sheer weight of their numbers and wealth. Older people hold on to their houses and change them to suit themselves, rather than trading up as they did when they were younger. The very advances in technology that make society conservative about buildings also make it technically easier to preserve old buildings and to update their services in subtle ways. And accommodating the disabled has proved, as it often does, to be prescient for the society as a whole. Wheelchair ramps and low-piled carpets, roomy bathrooms and lever-handled doors are a convenience for all of us as we slow down.

At the same time, formerly youthful environmentalists have come into power in business and government and are converting what used to be the shrill demands of outsiders into the law of the land. Health-conscious oldsters are supporting them. Buildings will be ever more stringently regulated in terms of their air quality, electromagnetic radiation, safety, energy use, and other issues yet to be identified.

Futurists always include technology as one of the driving forces in the future of nearly everything, from archaeology (DNA reconstitution) to literature (multimedia). What technologies will drive building construction and reconstruction? William Morris declared in 1892, "The subject of Material is clearly the foundation of architecture." These days architecture has no need to be innovative in materials science, since it can so easily borrow. One observer notes,

> Sir Norman Foster is well known for his ingenious early use of components and materials that have their origin in industries far removed from construction: solvent-welded PVC roofing derived from swimming pool liners, gaskets of neoprene developed originally for cable-jacketing, structural

glazing and glass fritting from the auto industry, superplastic aluminum and metalized fabrics from aerospace. Even Foster's presentation-drawing techniques are culled from aviation magazines.[10]

Factory manufacture has increased the use of synthetic materials and has brought high quality, low price, and a surprising degree of customization of components with it. A trade magazine reports that the use of plastics in construction is growing by 5 to 6 percent a year, reaching 43 pounds per thousand dollars of construction in 1989, having doubled in seven years.[11] Plastic is used in everything from planklike Durawood (made from recycled plastic containers) to soft bathtubs and flexible granite (no kidding; eighth-inch-thick granite-and-polymer veneer can be heat-formed to a fourteen-inch radius).

Such composite materials come from a long tradition. Renowned classical Roman buildings were made of concrete. Plywood furniture is found in ancient Egyptian tombs. Many a centuries-old decorative ornament in Europe is made of "composition"— animal-hide glue, burnt sienna, oil, whiting, and water. Modern ornamental materials include polyester, polyurethane foam, and fiberglass-reinforced gypsum. Fiberglass-reinforced concretes and polyesters are strong enough to imitate all sorts of traditional structural materials, and the drive for their innovation, ironically, has come from the boom in building preservation. The substitutes are lighter, cheaper, and easier to install than the decayed stone, plaster, or carved wood they replace, and they look exactly the same. The entire marble portico of an 1899 mansion in Lenox, Massachusetts, has been recreated, down to the last curlicue, out of fiberglass-reinforced polyester. A product called Cathedral Stone can be molded by hand to duplicate ancient masonry. Come back a week later when it's cured and you need a hammer and chisel. A little instant aging with muriatic acid makes it look just like the real stone next to it.

What are we to make of all this apparent fraudulence? As a boat lover I remember when fiberglass boats first came along in the 1950s, and everyone said they would never work, never sell, never last. Wrong on every count. Fiberglass boats are lighter and stronger than wood, more intricately shaped, and they endure negligent owners, which wood cannot, because they are immune to teredo worms, dry rot, and baking sun. Fiberglass never leaks; wood always does, top and bottom. And yet a magazine called *Woodenboat*, founded in 1969, became one of the all-time publishing successes through worshipping the virtues of wood in boats. Those virtues consist entirely of the aesthetics of tradition and the discipline of managing a short-lived material. I have owned and sold an excellent plastic boat and owned and kept a troublesome wooden boat. Why? The wood feels better, and I can fiddle with it. But if I really had to sail somewhere, I'd get fiberglass or steel.

I think that's what is happening with buildings. The ones that have to sail somewhere will be made of advanced materials. The ones that can be appreciated for the luxury and cost of touchable aesthetics (or whose labor costs are irrelevant) will stick with traditional materials. And most of the advanced stuff will ever more convincingly, and ironically, imitate the traditional, leading civilization into what Umberto Eco calls "hyperreality"—the realm of the exaggerated, proud, absolute fake.

---

10 Martin Pawley, "The Case for Uncreative Architecture," *Architectural Record* (Dec. 1992), p. 20.

11 Marylee MacDonald, *The Journal of Light Construction* (July 1990), p. 16.

12 One electronic device I can't wait for is "active noise control"—computerized noisemakers that duplicate a routine source of noise and rebroadcast the sound 180° out of phase with the original and thus effectively silence it. They cancel the din at the point of origin or at your ear, whichever is more convenient.

13 *Science* (17 Jan. 1992), p. 284.

14 With the passing of the Cold War, the vast defense industry is under pressure from government to swerve its technology toward what is called "dual use." Meaning: figure out some civilian applications or go out of business. Defense contractors are making calls on architects to push "smart materials."

15 John Ruskin, *The Seven Lamps of Architecture* (New York: Dover, 1849, 1880, 1989), p. 186.

Everyone expects electronic technology to transform building use. It already has, several times, but there are lots of false starts. The "intelligent buildings" fiasco of the early 1980s will probably be repeated in what the National Association of Home Builders wants to call "Smart Houses." This highly centralized, highly integrated, wire-based, skill-intensive approach attempts to install utter convenience and is more likely to install utter frustration. There is a standards battle going on between competing systems. People will be as baffled trying to program their house as they were trying to program the early video cassette players. I think that integrated electronics is indeed coming to the home, but by way of simple-minded wireless hobbyist devices, easy to buy one by one, easy to move around, easy to adjust to each other. Amateur house-hackers will have the garage door, the motion sensors, the phone, and the coffee maker all talking to one another by radio waves.[12]

A deeper revolution is evident in the title of a technical magazine founded in 1991, the *Journal of Intelligent Material Systems and Structure*. Self-sensing and self-healing materials are on the way. A researcher named Carolyn Dry, impatient with standard concrete ("brittle, porous and very dumb"), has developed a concrete laced with two kinds of fiber—one that detects when the steel reinforcing bar is corroding and releases an anti-corrosion chemical, and another that notices cracks and fills them with glue.[13]

The parallel advances of biotechnology and nanotechnology (molecular engineering) are bound to infect materials with either biological or nanocomputerized sensitivity in the coming decades. "Biobuildings" or "cognitive buildings" could brighten real-estate hopes for premium rents. Computerized weapons systems, these years, are graded by their designers in ascending order as: dumb, smart, brilliant, and moral. Materials systems will no doubt climb the same curve, and we can contemplate the coming of moral plastic.[14] In that case there will be added to the Low Road and the High Road an exquisitely tunable High Tech Road. It will be obvious how those buildings learn. They learn by paying attention.

But how might such buildings avoid making civilization even more pathologically ahistorical than it is? At times our era seems to enact Dante's warning, "Without hope we live in desire." Could our artifacts embrace enough future with enough liveliness to embody and radiate hope? Ruskin's plea was, "When we build, let us think that we build forever. Let it not be for present delight, nor for present use alone; let it be such work as our descendants will thank us for."[15] How exactly do good ancestors design?

Here's an exercise. Computer scientist Danny Hillis has proposed the making of "a large (think Stonehenge) mechanical clock, powered by seasonal temperature changes. It ticks once a year, bongs once a century, and the cuckoo comes out every millennium." The point is to have a charismatic object that helps people think long-term. No doubt a monastery of sorts should take care of the clock and its visitors, and also attend to other civilizational errands that operate at its pace. What kind of building would serve? No monuments, please. The design problem is to start a building which knows about centuries yet adeptly meets the needs and employs the tools of decades.

Time cures. In Tennyson's poem Sir Bedivere is mourning the passing of the Round Table, and the dying Arthur reassures him, "The old order changeth, yielding place to new,/ And God fulfills himself in many ways,/ Lest one good custom should corrupt the world." Instant-gratification, universal-standard buildings *are* corrupting. What is called for is the slow moral plastic of the "many ways" diverging, exploring, insidiously improving. Instead of discounting time, we can embrace and exploit time's depth. Evolutionary design is healthier than visionary design.

1868 - Near Virginia City in Nevada was the world's largest quartz mill, the Gould & Curry Silver Mining Company Reduction Works. In 1868 the silver was already playing out in the Comstock Lode. This photo was made by Timothy O'Sullivan for the US Geological Survey.

12:26 pm, 12 July 1979. Mark Klett.

1979 - Mark Klett rephotographed the identical view a century later (note stones in lower left of each photo) for *Second View*—reviewed on p. 229.

RECOMMENDED BIBLIOGRAPHY

Books for Time-kindly Buildings

THESE BOOKS are the best I came across (while perusing thousands) offering technique or inspiration for making, remaking, and appreciating buildings embedded in time. Some were major sources for my writing; many were not. They are the texts I would reach for if I was going to work on a building, rather than a mere book.

HOME

Renovation - (Michael W. Litchfield, Prentice-Hall, 1991) - There is no better book on fixing or improving a house. The book itself is a rehab—this is the second edition, 600 pages, 1,000 photos—a well-written, comprehensive work. It's aimed at do-it-yourselfers, but anyone about to hire a contractor should scan it first and have it handy during the work.

The Low-Maintenance House - (Gene Logsdon, Rodale, 1987) - What's shocking is that this is the only book on the subject. It's a competent job, a little quirky, but honest and helpful, full of lore that no one else tells you. I can't think of another book that would save so much money and aggravation in the long run.

Profits in Buying & Renovating Homes - (Lawrence Dworin, Craftsman, 1990; PO Box 6500, Carlsbad, CA 92008; 800 829-8123) - Nothing fancy, just solid advice on making money by improving houses. The author steers you toward basics and away from indulging fantasies, toward reality and away from half-smart real-estate schemes. A modest amount of this could turn any declining neighborhood

around. Banks, insurance companies, and city planners should just hand out books like the ones reviewed here.

The Walls Around Us - (David Owen, Villard, 1991) - As a former editor of the *Harvard Lampoon,* the author has wit to match his obsession to understand how houses really work. This book tells the (sometimes appalling) truth about wallboard, about lumber, about paint, about home-improvement projects that go on forever.

The Efficient House Sourcebook - (Robert Sardinsky, et al., Rocky Mountain Institute, 1992; Snowmass, CO 81654; 303 927-3851) - From Amory Lovins's esteemed energy research institute comes detailed recommendations of all the best books, suppliers, and organizations relating to having an energy-efficient house. Energy is the largest operating expense of a house. As someone pointed out years ago, storm windows (for example) are a safer investment with higher returns than anything on the stock market.

Small Spaces - (Azby Brown, Kodansha, 1993) - The book is an elegant jewelbox of ingenious design ideas for close-packing the elements of comfortable, even luxurious, living. The ungainly sprawl of oversized, wasteful houses can be cured right here.

Martha Stewart's New Old House - (Martha Stewart, Clarkson Potter, 1992) - This is the most thorough (and helpful) documenting ever of the restoration, renovation, decoration, and landscaping of a classy old house. The author is skillful at those tasks as well as showing and teaching the how-to.

Fine Homebuilding (magazine) - (Bimonthly, $29/year, PO Box 5066, Newtown, CT 06470) - A class act, the magazine always has the cutting edge techniques on quality homebuilding—equally useful for remodeling. The publisher, Taunton Press, also has a series of excellent homebuilding and woodworking books.

The Timber-Frame Home - (Tedd Benson, Taunton, 1988) - The beautiful, protected Structure of a timber-frame house is handily separate from Skin and Services, so it can last nearly forever. This is the top book on the subject.

The National Trust Manual of Housekeeping - (Hermione Sandwith & Sheila Stanton, Viking, 1991) - That's the British National Trust, reflecting on the upkeep of historic houses and their collections. If your house is a museum, or aspires to become one, here are the techniques to keep everything inside gorgeous and working.

Home - (Witold Rybczynski, Penguin, 1986) - The idea and form of "home" evolved dramatically over the centuries (and presumably still is). This deservedly popular book puts the present myth and practice in proper perspective.

Bonnettstown - (Andrew Bush, Abrams, 1989) - A superb photographer chronicles (in view-camera color) real life in a real house—a redolent old Irish manor house occupied by four elderly aristocrats. This book is the antidote to all those damned interior style books.

American Shelter - (Lester Walker, Overlook, 1981) - With children's-book clarity, the architect/author illustrates the full history of American houses, from pueblos to mobile homes, with every major style and period along the way. Unlike all other such books, he shows how the interiors work as well as how the exteriors look.

A Field Guide to American Houses - (Virginia and Lee McAlester, Knopf, 1984) - This richly-illustrated definitive work persuaded me that houses are a biological form, as alive and evolving as birds, because you can key out their species almost as precisely (and with the same delight) as you can with birds. Before building or buying a house, I'd spend some time with this book, to learn the traits and range of what's out there.

DESIGN

A Pattern Language - (Christopher Alexander, et al., Oxford, 1977) - The one book that everyone dealing with buildings (or towns) should have. I know people who have stayed up half the night reading it like a novel they can't put down. The organization of the information is radical, and impeccable—here are the details that make buildings *work*

(people prefer small-paned windows, shallow balconies don't get used, doors work best on the diagonal in rooms), sequenced and nested and cross-referenced in a way that makes a larger sense of how buildings come to life as a whole. It is a tour-de-force.

The Timeless Way of Building - (Christopher Alexander, Oxford, 1979) - Behind *A Pattern Language* is a philosophy about design and buildings of great depth, savvy, and subtlety. In a sense, this is a treatise on evolutionary design. In another sense, it is about the ethics of design.

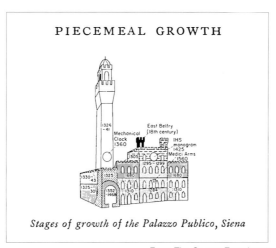

From *The Oregon Experiment*

The Oregon Experiment - (Christopher Alexander, et al., Oxford, 1975) - Principles of community planning (tried out at the University of Oregon) explored here have equal application in individual buildings—particularly the concept of "piecemeal growth:" periodically improving artifacts at every scale simultaneously.

A New Theory of Urban Design - (Christopher Alexander, et al., Oxford, 1987) - The most succinct and recent statement of

Alexander's still-evolving design philosophy. It makes a persuasive case for *accelerated* evolutionary design.

3D Home Architect (software) - (Brøderbund, 1994) - Architecture without architects. There's no better way to try out space plans, room design, and building configurations than with this intuitively adept software (the best as of 1994). The program prevents you from doing impossible things, offers recommendations, and provides 3D views of what you design, along with precise dimensions and materials lists. It's CAD (computer-aided design) for amateurs. It won't do finished drawings; for that you still need an architect.

The Art of the Long View - (Peter Schwartz, Doubleday, 1991) - Until there is a book on scenario planning specifically for buildings, this is the best introduction to the field. By being imaginative and conservative about the future, buildings can remain exceptionally alert to the on-going present.

Problem Seeking - (William Peña, et al., AIA, 1987) - The standard primer on programming—discerning the needs of the expected users of a building in a structured way.

Post-Occupancy Evaluation - (Wolfgang Preiser, et al., Van Nostrand Reinhold, 1988) - The standard primer on POEs—discerning what went wrong (and right) with a building in a structured way.

The Occupier's View - (Vail Williams, 1990, £50 from Vail Williams, 43 High Street, Fareham, Hampshire, PO16 7BQ, England) - The scalding product of post-occupancy evaluation of fifty-eight new business buildings near London. I would not set about building or remodeling a business building without studying this report first. Then I'd go talk to nearby facilities managers to learn about local problems of the same sort. The report stands as an indictment of contemporary design, construction, and real estate practices.

The Changing Workplace - (Francis Duffy, Phaidon, 1992) - This is an anthology of the writings of Frank Duffy, the architect who noticed and codified the layering of change in buildings. His primary focus is office environments, but his insights often apply much wider. All the articles and essays collected here are enlivened by Duffy's recent comments on them in the margins.

The Total Workplace - (Franklin Becker, Van Nostrand Reinhold, 1990) - Now that there is an enlightened textbook like this for facilities managers of "elastic organizations," other building professionals may want to investigate the design implications.

The Design Process - (Ellen Shoshkes, Whitney, 1989) - Here are nine fairly unvarnished case studies of architectural design in action, both for buildings and complexes, including one by William Rawn's office and one by Robert Venturi's office. Practitioners may want to see how the competition works; potential clients should want to see what they're in for.

The Architectural Theory of Viollet-le-Duc - (M. F. Hearn, editor, MIT, 1990) - The founder and inspirer of the preservation movement and the Modern movement, Viollet-le-Duc (1814-1879) still surpasses his followers with penetrating good sense. This annotated sampler should beguile professionals to track down his original works.

The Image of the Architect - (Andrew Saint, Yale, 1983) - How did the architecture profession paint itself into its present corner—

obsessed with image, negligent of process, prone to arrogance? British architectural historian Saint has the whole sordid, amusing history.

PRESERVATION

Historic Preservation (magazine) - (Bimonthly, comes with National Trust membership, $20/year, 1785 Massachusetts Ave., NW, Washington, DC 20078) - America's National Trust for Historic Preservation is singularly well organized to nurture anyone's interest in old buildings. It offers a variety of publications, conferences, and extensive networking access to state and local preservation groups. The magazine does a nice job of feeding the obsession and tracking the growth of preservation philosophy.

All About Old Buildings - (Diane Maddex, editor, Preservation Press, 1985) - Though out of print, this extraordinary compendium is worth finding in a library or local preservation office, because it offers the most thorough and imaginative introduction to the full scope and depth of preservation activities. A more recent, more limited, more available version is **Landmark Yellow Pages** (Preservation Press, 1990).

Illustrated Guidelines for Rehabilitating Historic Buildings - (National Park Service, 1992, $8 from Superintendent of Documents, Government Printing Office, Washington, DC 20402-9325) - The still-evolving core doctrine of American preservation is expressed in the celebrated "Secretary of the Interior's Standards for Rehabilitation," here spelled out in illustrated detail. It is the equivalent of a building code for old buildings—sensible guidance, required if you want to meet the

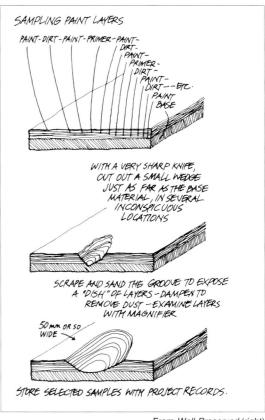

From *Well-Preserved* (right)

standard for official approval or financial assistance.

Preservation Briefs - (National Park Service) - These small, dense booklets offer the best technical advice on specific subjects such as repointing masonry, preserving adobe, repairing wood windows, preserving terra cotta, keeping old barns alive, and so on. They and other resources are listed in the **Catalog of Historic Preservation Publications** (Government Printing Office, above; or contact 202 343-9578 for assistance).

Care for Old Houses - (Pamela Cunnington,

London: A & C Black, 1991) - Much the best book on the subject in America or Britain, both for technical detail and aesthetic guidance.

Historic Preservation - (James Marston Fitch, Univ. of Virginia, 1990) - The founder of preservation as an academic discipline in America, Fitch has written and updated the outstanding standard text. Its scope is refreshingly global.

Well-Preserved - (Mark Fram, 1988, Boston Mills Press, 132 Main Street, Erin, Ontario, Canada N0B 1T0) - This Canadian labor of love offers exceptionally thorough and practical advice on every aspect of building preservation.

New Orleans Architecture - (Pelican, Gretna LA, 1971-1989) - Of the growing library of city compendia of landmark buildings, the two most exemplary are the New Orleans volumes and **Jacksonville's Architectural Heritage** (Univ. of North Florida, 1989). The Jacksonville book gains depth by printing both historical and contemporary photos of its buildings. The New Orleans volumes (up to seven so far) are unmatched for quantity and quality of coverage. They include the beautiful notarial watercolor drawings made in the 19th century—one is on the cover of *How Buildings Learn.*

CITY

The Death and Life of Great American Cities - (Jane Jacobs, Modern Library, 1961, 1993) - A classic, now recognized as such in a fine Modern Library edition. With writing of literary quality, Jacobs proved that the vitality of cities comes from density and diversity—both anathema to simplistic schemes of city planners.

Edge City - (Joel Garreau, Doubleday, 1988) - Garreau's book is just as embarrassing to

architects and city planners as Jacobs's was, because the real action (this time in new peripheral cities) has once again eluded their notice and theory. Garreau, a demographer and *Washington Post* journalist, reveals more of present-day American reality than any other book I can think of. (Garreau edits an expensive newsletter to keep up with the rapid change in edge cities, from PO Box 1145, Warrenton, VA 22186.)

Downtown, Inc. - (Bernard J. Frieden & Lynne B. Sagalyn, MIT, 1989) - Forced to compete with suburban shopping malls, downtowns responded by imitating them, while enhancing the unique advantages of downtown centrality, age, and direct access to city government. A scholarly but readable study.

The Next American Metropolis - (Peter Calthorpe, Princeton Univ., 1993) - One of the new generation of "neotraditional" town planners spells out the elements of new communities built dense enough to be congenially walkable and to be economically served by mass transit.

City - (William H. Whyte, Doubleday, 1988) - In this compelling book Whyte performs the kind of study I would like to see made of buildings—insightful, prolonged observation of actual behavior, followed by completely revised design principles in service of that behavior. The book is about what makes downtown streets exciting.

Built for Change - (Anne Vernez Moudon, MIT, 1986) - Extensive analysis of a large area of San Francisco Victorian row houses showed that they and their neighborhoods owe their remarkable resilience to just a few major patterns—including small lot size, individual ownership, and easily adaptable room layouts.

What Time Is This Place? - (Kevin Lynch,

MIT, 1972) - A famed urban theorist celebrates how time gives depth to cities, and how design can honor and serve that.

Barmi - (Xavier Hernandez, et al., Houghton Mifflin, 1990) - It's a children's book, naturally, since adult books are not permitted to be so delightful. The subject of the huge, detailed illustrations is a Mediterranean city from 500 BC to the present. This is the way to think about buildings and change—in terms of centuries.

VERNACULAR

Wheel Estate - (Allan D. Wallis, Oxford, 1991) - Why do the most important and interesting subjects get so few books about them? In terms of innovation and quantity, trailers and mobile homes led the way in American housing in the later 20th century. Here was the subversive arrival of factory-made housing. Wallis scooped everybody with his excellent book.

The Malay House - (Lim Jee Yuan, 1987, Institut Masyarakat, 87 Cantonment Road, 10250 Pulau Pinang, Malaysia) - I've seen no better examination of how a successful building type works. This is a brilliant and beautiful book on a superbly adaptive vernacular house form.

Big House, Little House, Back House, Barn - (Thomas C. Hubka, University Press of New England, 1984) - This lucid work dissects the history and functionality of the famed New England "connected" farms, including even diagrams of how woman's work and man's work was distributed spatially.

Japanese Homes and Their Surroundings - (Edward S. Morse, Dover, 1886, 1961) - A 19th-century masterpiece, this exquisite little book conveys the genius of traditional Japanese houses in detail never surpassed. The quiet influence of this book keeps being renewed decade after decade.

Early Nantucket and Its Whale Houses - (Henry Chandlee Forman, Nantucket: Mill Hill, 1966) - An exhaustive, fun investigation of the tiny colonial houses of Siasconsett, Nantucket, and how they grew.

Discovering the Vernacular Landscape - (J. B. Jackson, Yale, 1984) - Highly original, incisive essays from the modern prophet of American vernacular studies. A second set, equally valuable, is in **The Necessity for Ruins** (Univ. of Massachusetts, 1980).

HISTORY

Jefferson and Monticello - (Jack McLaughlin, Holt, 1988) - A house biography based on one of the great biography-houses. A man's vision and his life were fully embodied, drawing by drawing, brick by brick, in Monticello—in every sense, a life work.

The President's House - (William Seale, Abrams, 1986) - The White House may be the best-documented building in the world; it is certainly the best in America. As George Washington expected (and designed for), it has borne extensive and consequential change. Seale is an architect, an architectural historian, a restoration consultant, and an excellent writer. Through his two-volume version of the White House, we see American history unfold.

The House - (The Duchess of Devonshire, London: Macmillan, 1982) - I took Chatsworth as my emblematic High Road building because in Deborah Devonshire's writing, it is as if the house itself were speaking—remembering favorite dukes, telling fine old stories of historic

occasions and ambitious follies and eccentric successes, thanking staff and caretakers, complaining knowledgeably about the sheer hassle of keeping up with such a house. The book is ably matched by her later work, **The Estate** (London: Macmillan, 1990), about maintaining Chatsworth as a splendid place on earth.

Ancient English Houses - (Christopher Simon Sykes, London: Chatto & Windus, 1988) - This is far my favorite book on great old English houses—1240 to 1612 (Chatsworth is too recent to qualify). Since Sykes did both the photography and the writing, the work has an integrity lacking in many picture books, and he got into some houses that no one else does. These are High Road buildings at their finest.

English Cathedrals: The Forgotten Centuries - (Gerald Cobb, London: Thames & Hudson, 1980) - England's medieval cathedrals kept right on changing dramatically through the 17th, 18th, 19th, and 20th centuries, but everyone wanted to forget about that until Cobb proved that cathedrals were just as malleable as any other kind of building. The book is an exciting read, with rich graphics.

The Master Masons of Chartres - (John James, West Grinstead, 1982, 1990) - Stone by stone, James unravels the mystery of the construction of one of the world's most cherished buildings. The thing grew almost biologically—a mess up close, sublime as a whole. The more extensive version of James's research is **The Contractors of Chartres** (1981, Wyong, Australia: Mandorla).

The Plan of St. Gall - (Walter Horn & Ernest Born, Univ. of California, 1979) - Another detective story is Horn & Born's decrypting of history's most intriguing town plan—for a 9th-century Benedictine monastery complex. The

original three-volume edition of this work is simply magnificent. The paperback condensation (**The Plan of St. Gall in Brief**, Univ. of California, 1982) has much of the substance and some of the glory.

REPHOTOGRAPHY

I was so reliant on rephotography by others that I would like to make my own photographs easily available, hoping that some photographers may continue some of the sequences in this book. Accordingly, any of my prints are available for $25 apiece (includes print from original negative and permission to reprint), plus shipping, from Stewart Brand, Global Business Network, PO Box 8395, Emeryville, CA 94662; fax 510 547-8510.

Cityscapes of Boston - (Robert Campbell & Peter Vanderwarker, Houghton Mifflin, 1992) - Thanks to Campbell's insightful commentary, this is best book of building rephotography so far. You can see Boston's urban intelligence and obliviousness alive in time. Photographer Vanderwarker is careful to duplicate the feel as well as the look of earlier images in his rephotographs.

Boston Then and Now - (Peter Vanderwarker, Dover, 1982) - This earlier work by Vanderwarker has no repeats from the *Cityscapes* book. It is one of the fascinating series of paperback rephotography books from Dover Publications in New York.

Washington, DC, Then and Now - (Charles Suddarth Kelly, Dover, 1984) - Since Kelly collects as well as makes photographs of Washington, he has the best multiple-image

ca. 1760 - Piranesi's etching of the Temple of Saturn in Rome. To its immediate left is the Arch of Septimius Severus (203 AD), with the Baroque church of S. Martino al Monti (remodeled 1640) in the background.

ca. 1970 - Herschel Levit's rephoto for *Views of Rome Then and Now* (see below).

sequences of streets and buildings changing and changing and changing.

Nantucket Yesterday and Today - (John W. McCalley, Dover, 1981) - I found this book particularly interesting because it focuses on a small town and its environs instead of a city— and a historically-minded small town at that.

Philadelphia Then and Now - (Kenneth Finkel & Susan Oyama, Dover, 1988) - Finkel tells good gossipy stories to go along with Oyama's excellent photos.

New York Then and Now - (Edward B. Watson & Edmund V. Gillon, Dover, 1976) - New York is more interested in real estate and less interested in history than any other East Coast city, and it shows.

Views of Rome Then and Now - (Piranesi & Herschel Levit, Dover, 1976) - You can't ask for a more observant predecessor than Giovanni Battista Piranesi, whose 18th-century etchings of Rome prove to be photographically precise. The book is large-format to make the most of Piranesi's extraordinary depiction of detail. The current photos show what city planner Benito Mussolini did to Rome.

Santa Fe Then and Now - (Sheila Morand, Santa Fe: Sunstone, 1984) - It is amazing to watch this town make itself over in an imaginary image of its past—ersatz so complete that it creates its own reality. It is the selective, revisionist memory of nostalgia made physical. Thousands love the result; me too.

San Francisciana: Photographs of the Cliff House - (Marilyn Blaisdell, San Francisco: Blaisdell, 1985) - The best photo-series I've seen of a spectacular site attracting and shedding building after building.

Stopping Time - (Peter Goin, Univ. of New Mexico, 1992) - Changes in a whole region— the Lake Tahoe area in California—are documented in this careful, somewhat art-oriented book. As with *Second View* (below), which it derives from, you see the impact of human development on a wild environment.

Second View - (Mark Klett & Ellen Manchester, et al., Univ. of New Mexico, 1984) - Rephotography became a creative discipline with the publication of this book. It gives the technique of how to rephotograph accurately and with perception, and it is an inspiring exemplar with its striking 19th-century landscape photos of the American West juxtaposed with recent view-camera photographs of the identical places.

Index